International Conference on Materials and Intelligent Manufacturing (ICMIM 2017)

IOP Conference Series: Materials Science and Engineering
Volume 244

Singapore
21 – 23 August 2017

ISBN: 978-1-5108-4999-0
ISSN: 1757-8981

Printed from e-media with permission by:

Curran Associates, Inc.
57 Morehouse Lane
Red Hook, NY 12571

Some format issues inherent in the e-media version may also appear in this print version.

Printed by Curran Associates, Inc. (2017)

For permission requests, please contact the Institute of Physics
at the address below.

Institute of Physics
Dirac House, Temple Back
Bristol BS1 6BE UK

Phone: 44 1 17 929 7481
Fax: 44 1 17 920 0979

techtracking@iop.org

Additional copies of this publication are available from:

Curran Associates, Inc.
57 Morehouse Lane
Red Hook, NY 12571 USA
Phone: 845-758-0400
Fax: 845-758-2633
Email: curran@proceedings.com
Web: www.proceedings.com

Table of contents

Volume 244

2017 International Conference on Materials and Intelligent Manufacturing (ICMIM 2017)
21–23 August 2017, National University of Singapore (NUS), Singapore

Accepted papers received: 18 September 2017
Published online: 16 October 2017

Preface

011001
OPEN ACCESS
2017 International Conference on Materials and Intelligent Manufacturing (ICMIM 2017)

011002
OPEN ACCESS
Organizing Committee

011003
OPEN ACCESS
Peer Review Statement

Materials Processing Engineering

012001
OPEN ACCESS
Sustainability assessment of shielded metal arc welding (SMAW) process

Ibrahim Alkahla and Salman Pervaiz.....1

012002
OPEN ACCESS
Investigation on burrs in micro milling of stainless steel 310S

Yuan Meixia, Liu Shaonan, Xue Hongxin, Tang Boyan and Zhao Linlin.....9

012003
OPEN ACCESS
Prediction of multi performance characteristics of wire EDM process using grey ANFIS

Somasundaram Kumanan and Anish Nair.....16

012004
OPEN ACCESS
Study of wear performance of deep drawing tooling

Vishal G Naranje, Ram Karthikeyan and Vipin Nair.....24

012005
OPEN ACCESS
Influence of cryogenic cooling rate on mechanical properties of tool steels

G Mazor, I Ladizhensky and A Shapiro.....32

012006
OPEN ACCESS
Comparison of tool life and surface roughness with MQL, flood cooling, and dry cutting conditions with P20 and D2 steel

S W M A I Senevirathne and H K G Punchihewa.....39

Polymers and Composites

012007
OPEN ACCESS
Effect of Commercial SiO_2 and SiO_2 from rice husk ash loading on biodegradation of Poly (lactic acid) and crosslinked Poly (lactic acid)

C Prapruddivongs, M Apichartsitporn and T Wongpreedee.....47

012008
OPEN ACCESS
Thermal properties of biopolyol from oil palm fruit fibre (OPFF) using solvolysis liquefaction technique

Shaharuddin Kormin, Anika Zafiah M. Rus and M. Shafiq M. Azahari.....52

012009
OPEN ACCESS
Acoustic properties of polymer foam composites blended with different percentage loadings of natural fiber

M. Shafiq M. Azahari, Anika Zafiah M. Rus, Shaharuddin Kormin and M. Taufiq Zaliran.....58

012010
OPEN ACCESS
A novel technique for producing conductive polyurethane nanofibrous membrane for flexible electronics applications

A Shaker, Ahmed H Hassanin, N M Shaalan, M A Hassan and Ahmed Abd El-Moneim.....64

012011
OPEN ACCESS
Preliminary investigation of fabrication composite structures by using shape memory alloys

W Klein and O Dudek.....71

012012
OPEN ACCESS
Wear properties of 10 vol.% silicon carbide particulate-reinforced aluminum composite fabricated by powder injection molding

T Patcharawit, A Ngeekoh and N Chuankrekkul.....77

012013
OPEN ACCESS
Investigation of the fatigue crack propagation behaviour in the Al alloy/Hybrid MMC Bi-layer material

AKM Asif Iqbal and Yoshio Arai.....85

Materials Physics and Computational Materials Science

012014
OPEN ACCESS
Adsorption and desorption behavior of CO on S passivated 4H-SiC by electric field

Zhengxin Yan, Anhong Hou, Yongsheng Wang and Shaohua Zhang.....91

012015
OPEN ACCESS
Analyzing of a 2-D Magnet Array with Hexagon Magnet based on superposition

Guangdou Liu, Xin Zhang, Xingping Xu and Xiaoyi Liu.....97

012016
OPEN ACCESS
Constitutive modelling of elastomer/graphene platelet nanocomposites

Amir A. Abdelsalam, Sherif Araby, M. A. Hassan and A. A. El-Moneim.....102

012017
OPEN ACCESS
An improved edge detection approach and its application in defect detection

Vilas H Gaidhane, Navdeep, Asha Rani and Vijander Singh.....110

012018
OPEN ACCESS
Thermal deformation analysis of Spliced high energy laser emission window

Xutang Liu, Diansheng Cao and Guangdou Liu.....117

012019
OPEN ACCESS
Motion performance analysis and control mode design for the cross slide

Dong Xie, Jianqu Zhu, Min Wang, Zhengzhong Li and Feng Wang.....123

012020
OPEN ACCESS
Social aspects of automation: Some critical insights

Ibrahim Nouzil, Ali Raza and Salman Pervaiz.....130

012021
OPEN ACCESS
Study on vacuum packaging reliability of micromachined quartz tuning fork gyroscopes

Maoyan Fan and Lifang Zhang.....138

012022
OPEN ACCESS
Computer aided system for parametric design of combination die

Vishal G Naranje, H M A Hussein and S Kumar.....147

Materials Characterization and Application

012023
OPEN ACCESS
Buildup factor and mechanical properties of high-density cement mixed with crumb rubber and prompt gamma ray study

P Aim-O, D Wongsawaeng, S Tancharakorn and M Sophon.....154

012024
OPEN ACCESS
Enhancement of mechanical properties of porous aluminum by silica sand particles

S Khamsuk, A Joosawat, N Panomtang and K Wongtimnoi.....161

012025
OPEN ACCESS
Study on service performance of 880 MPa-grade and 980 MPa-grade rail steels

M Zhu, G Xu, J H Zhou, R M Wang and X L Gan.....167

012026
OPEN ACCESS
The morphologies of oxide scale in a Si-containing steel

M X Zhou, G Xu and Q Yuan.....173

012027
OPEN ACCESS
Industrial wastewater treatment using higher aquatic vegetation in the former mining company of the Far Eastern Federal district

L. T. Krupskaya, V. P. Zvereva, K. E. Gula, L. P. Gul', D. A. Golubev and M. Yu. Filatova.....179

IOP Conf. Series: Materials Science and Engineering **244** (2017) 011001 doi:10.1088/1757-899X/244/1/011001

PREFACE

It is our great pleasure to welcome you to 2017 International Conference on Materials and Intelligent Manufacturing (ICMIM 2017) which was held during 21-23 August, 2017 in Singapore. ICMIM 2017 is dedicated to issues related to Materials and Intelligent Manufacturing.

The major goal and feature of the conference is to bring academic scientists, engineers, industry researchers together to exchange and share their experiences and research results, and discuss the practical challenges encountered and the solutions adopted. Professors from China, Japan and Singapore are invited to deliver keynote speeches regarding latest information in their respective expertise areas. It will be a golden opportunity for the students, researchers and engineers to interact with the experts and specialists to get their advice or consultation on technical matters, sales and marketing strategies.

This proceedings present a selection from papers submitted to the conference from universities, research institutes and industries. All of the papers were subjected to peer-review by conference committee members and international reviewers. The papers selected depended on their quality and their relevancy to the conference. The volume tends to present to the readers the recent advances in the field of materials processing technology and materials science, design and manufacturing systems, advanced design technology, etc.

We would like to thank all the authors who have contributed to this volume and also to the organizing committee, reviewers, speakers, chairpersons, sponsors and all the conference participants for their support to ICMIM 2017.

Prof. Jun Ding
National University of Singapore, Singapore
August 30, 2017

IOP Conf. Series: Materials Science and Engineering **244** (2017) 011002 doi:10.1088/1757-899X/244/1/011002

Organizing Committee

Conference Chairs

Prof. Jun Ding, National University of Singapore, Singapore

Prof. Lei Jiang, Chinese Academy of Sciences, Beijing, P.R.China

Technical Program Committee Chairs

Prof. Yoshihiro Sugi, Gifu University, Japan

Assoc. Prof. Pan Jisheng, Institute of Materials Research and Engineering/ National University of Singapore (NUS), Singapore

Organizing Chair

Ms. Yang Dan, Hong Kong Chemical, Biological & Environmental Engineering Society, Hong Kong

Publication Chair

Ms. Tammy Luo, Chemistry and Materials Society, Hong Kong Chemical, Biological & Environmental Engineering Society, Hong Kong (CMS-HKCBEES)

Technical Program Committee

Prof. Wankei Wan, Western University, Canada

Prof. Flavio Horowitz, UFRGS (Federal University of Rio Grande do Sul), Brazil

Prof. SungCheal Moon, Korea Institute of Materials Science (KIMS), Republic of Korea

Prof. Euijoon Yoon, Seoul National University, Korea

Prof. Heung Nam Han, Seoul National University, South Korea

Prof. Chang-Sik Ha, Pusan National University, South Korea

Prof. Seung R Paik, Seoul National University, South Korea

Prof. Xu Guang, Wuhan University of Science and Technology, China/ Research and Development Center, WISCO, China

Prof. Wan Jefrey Basirun, University of Malaya, Malaysia

Prof. Ahmed Abd El-Moneim, Egypt-Japan University of Science and Technology, Egypt

Prof. Manoj Kumar Singh, CSIR-Central Institute of Mining & Fuel Research, India

Assoc. Prof. Togay Ozbakkaloglu, The University of Adelaide, Australia

Assoc. Prof. Mohammad Nahid Siddiqui, King Fahd University of Petroleum & Minerals, Saudi Arabia

Assoc. Prof. Anika Zafiah Mohd Rus, Universiti Tun Hussein Onn Malaysia, Malaysia

Assoc. Prof. A. Nirmala Grace, VIT University, India

Dr. Mazeyar Parvinzadeh Gashti, Chief Scientist, PRE Labs Inc., Canada

Dr. Alain Kusmoko, University of Wollongong, Australia

Dr. Wojciech Klein, Silesian University of Technology, Poland

Dr. Salman Pervaiz, Rochester Insititute of Technology- Dubai Campus, Dubai, UAE

Dr. Fan Maoyan, Yuxi Normal University, China

Dr. Kok Boon Chong, Sunway University, Malaysia

Dr. Oh Pei Ching, Universiti Teknologi PETRONAS Malaysia, Malaysia

Dr. Vimal S. Joshi, Sardar Patel University, India

IOP Conf. Series: Materials Science and Engineering **244** (2017) 011003 doi:10.1088/1757-899X/244/1/011003

Peer review statement

All papers published in this volume of *IOP Conference Series: Materials Science and Engineering* have been peer reviewed through processes administered by the proceedings Editors. Reviews were conducted by expert referees to the professional and scientific standards expected of a proceedings journal published by IOP Publishing.

IOP Conf. Series: Materials Science and Engineering **244** (2017) 012001 doi:10.1088/1757-899X/244/1/012001

Sustainability assessment of shielded metal arc welding (SMAW) process

Ibrahim Alkahla [1] **and Salman Pervaiz** [1]

[1]Department of Mechanical and Industrial Engineering
Rochester Institute of Technology – Dubai Campus
P.O. Box 341055 Dubai, United Arab Emirates
Email: sxpcad@rit.edu

Abstract. Shielded metal arc welding (SMAW) process is one of the most commonly employed material joining processes utilized in the various industrial sectors such as marine, ship-building, automotive, aerospace, construction and petrochemicals etc. The increasing pressure on manufacturing sector wants the welding process to be sustainable in nature. The SMAW process incorporates several types of inputs and output streams. The sustainability concerns associated with SMAW process are linked with the various input and output streams such as electrical energy requirement, input material consumptions, slag formation, fumes emission and hazardous working conditions associated with the human health and occupational safety. To enhance the environmental performance of the SMAW welding process, there is a need to characterize the sustainability for the SMAW process under the broad framework of sustainability. Most of the available literature focuses on the technical and economic aspects of the welding process, however the environmental and social aspects are rarely addressed. The study reviews SMAW process with respect to the triple bottom line (economic, environmental and social) sustainability approach. Finally, the study concluded recommendations towards achieving economical and sustainable SMAW welding process.

1. Introduction

Welding process joins materials by achieving localized coalescence under the action of heat, pressure or both. The localized coalescence help the faying surfaces to fuse in to each other and make single unit [1]. Welding is mainly utilized to fabricate different constructional structures by means of joining metallic elements. Welding process is available in all small or large scale workshops, where it is employed to repair the problematic parts as well. A good portion of production sector is based on the tasks related to the applications of different joining techniques. Welding has an advantage that it has been practiced in the space as well. It the fabrication process that has practical application in air, water and space. Welding is employed in the different industrial sectors such as automotive, aerospace, ship-building, construction, nuclear, , energy - power generation, electronics, house-hold appliances, petrochemicals and machinary fabrication etc. [1, 2]. In the form of different governmental rules and regulations, production sector is under high pressure to improve the environmental performance of different manufacturing operations [3, 4]. Besides the strict environmental legislations, production sector has also realized that improving environmental perfromance can bring long term economic benefits and better survival in the market [5].

The concept of sustainable development (SD) is gaining popularity in the manufacturing sector as well to explore the possibility of achieving development and imporved environmental performance simultaneously [6]. The idea of sustainable development (SD) deals with progressing the human development while sustaining the environment at the same time [7, 8]. Improving the environmental performance of a manufacturing process means reducing input energy consumption, reducing hazardious input and output material flows, improving work conditions, reducing risks related to the occupational safety and optimizing related costs etc [9]. Several researchers have focused their research work towards the betterment of the welding process in terms of increasing process efficiency, cost optimization, social aspects and environmental concerns. There is a need to characterize sustainability for the welding processes under the collective framework of different economic, environmental and social aspects together. As per literature, the sustainability assessment method where all economic, environmental and social aspects are collectively considered is termed as triple bottom line (TBL) or three pillar approach [10]. This study reviews and characterize sustianbilty for the SMAW process using TBL methodology. The study was helpful in providing a collective sustianbility overview using economic, environmental and social aspects. In the end, the study also recommends future directions for the research work to make SMAW process more environment friendly in nature.

As addressed earlier, welding operation is one of the most commonly used joining process in the metal fabrication industry. Figure 1 represents the input output inventory for the SMAW process. To perform the SMAW operation, electric current flows through the electrode and the workpiece material in the form of electric arc. The flow of electricity from electrode to workpiece generates heat on both sides and droplets of molten metal transfers to the molten puddle to fuses them together. Once the electrode moves away from the processing region the welding bead gets solidyifies and welding operation completes[11, 12]. The electrode is generally coated with the flux, which metals and decomposes during the welding operation and forms a kind of protecting shielding layer on the top of the weld bead to protect it from atomospheric oxygen. The SMAW process is famous in industry as it is very cheaper to operate, can weld most of the commonly used metals and alloys and easy to port in harsh environments etc. At the same time there are some limitations associated with the SMAW process such as slag has to be removed and frequent changing of electrodes makes the process slower in nature [13].

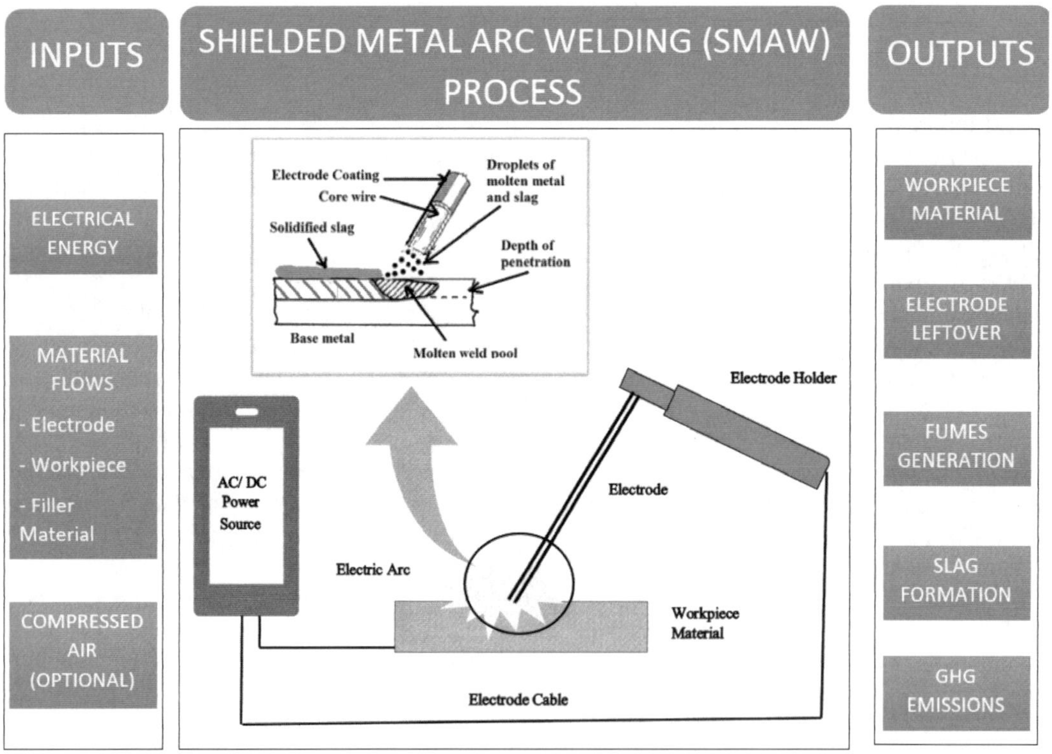

Figure 1. Input and output inventory for the SMAW process ([14,15])

IOP Conf. Series: Materials Science and Engineering **244** (2017) 012001 doi:10.1088/1757-899X/244/1/012001

2. Establishing framework using triple bottom line methodology

The idea of sustainable development is mainly defined as meeting the needs of current generation without compromising the needs of future generations [16]. Sustainable development plays a key role towards the sustainance of all living ecosystems, as it is not only for the human beings it also includes all living beings in the environment. When analysing systainability for certain operation, it is important to incorporate all three environmental, economic and social aspects simultaneously. Triple bottom line (TBL) or three pillar of sustainability approach points out at the integration of all three aspects [17–20]. In order to properly address all three components of sustainable development, it is very important to under the input and output inventory involved in the SMAW process. For the SMAW process the input streams are filler material, electrode and energy consumption. However, the output streams are the final workpiece material, slag and fumes. Figure 2 establishes the TBL based Venn diagram for the SMAW process. Each aspect of TBL sustainability approach is discussed in detail under the light of available literature.

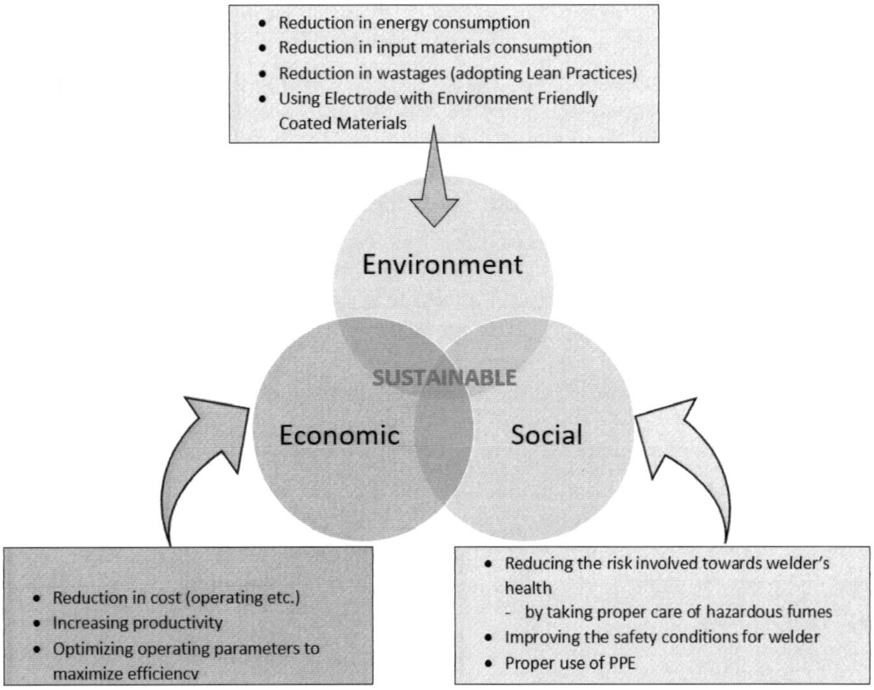

Figure 2. Triple bottom line (TBL) [21] approach for the SMAW process

3. Discussion

The study provided a framework to characterize sustainability for the most commonly employed welding operation namely SMAW process. Most of the studies available in the state-of-the-art are focused on the technical performance and social aspects of the SMAW operation. However, the presented paper incorporates all social, economic and environmental aspects together by using the fundamental sustainability approach of triple bottom line (TBL). The study is very helpful to see the broader perspective of SMAW process using all three dimensions together as a whole. It can be observed very clearly that the SMAW process has economic advantage over the other welding processes. In addition to the economic gain, SMAW process is also favoured in the industry due to its ease in transportability, which makes its promising for the harsh environments. When environmental impacts of different welding operations were investigated, it has been observed that highest environmental impact was found for manual SMAW process. This reveals the potential of automation in the welding operation. The automated version of arc welding operation can be utilized to reduce the environmental impact of the manual arc welding operation.

The concept of lean manufacturing deals with the minimization of waste throughout the manufacturing process [22]. The result of lean manufacturing based practices generally also results in lowering the environmental burden of the manufacturing process. Similarly, to achieve environment friendly welding operation, implementation of lean practices can be very helpful. Literature reveals [23], [24] some of the lean practices related to the welding operation, such as improving product quality by achieving zero defect, efficient use of all valuable resources, just-in-time material flows, process continuous improvement and introducing flexibility in manufacturing to maximize resource efficiency. In order to extend the discussion, each circle (economic, environmental and social) of TBL approach was further explored in the light of available literature.

3.1 Economic Aspects

It has been reported that annual electricity consumption of welding is \$15 million in the US and \$99 million globally [22]. Several researchers have focused their work to study the economic aspect of SMAW process. In addition welding operation is a costly process because it involves several types of resources. Performing parametric optimization using different experimental setups would be very expensive solution. Therefore, different approaches including Taguchi method, Genetic Algorithms and other optimizational techniques are very cost effective. Vimal et al. [23] utilized the graph theory approach to model different sustainable measures for the assessment of sustainable strategies. The study incorporated two sustainable strategies of employee skill training and waste reduction. The study revealed that waste reduction through the proposed optimization method bring both economic and environmental benefits. Jármai et al. [24] also conducted a cost optimization study for cylindrical orthogonally stiffened shell member. The study considered the cost function based on the raw materials, forming, welding and painting operations. The study utilized novel mathematical optimizations approaches namely leap-frog – LFOPC, Dynamic-Q, and particle swarm – PSO. The optimal results revealed remarkable margin of material and manufacturing related cost saving by reducing the shell thickness.

Patrick and Newell [25] explored the idea of cost reduction for the arc welding operation. The study revealed different labour and material related costs for the welding operation. It has been observed that costs related to material consumables is only 10 – 20% of overall welding cost, whereas the remaining 80 – 85% cost is related to the labour and overheads. Increase in the productivity can reduce the labour man-hour and decrease the associated cost. Figure 3 represents different deposition rates for various arc welding operations. It can be observed that maximum number of man-hours were reported for the SMAW process. The study suggested to replace SMAW process with another FCAW process to improve overall productivity.

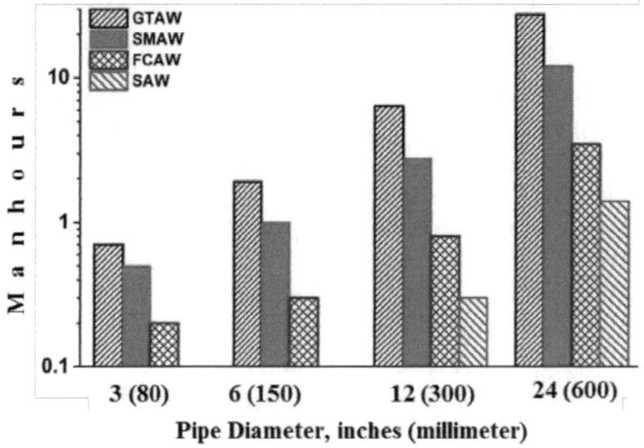

Figure 3. Deposition rate for Gas Tungsten Arc Welding (GTAW), Submerged Arc Welding (SAW), Shielded Metal Arc Welding (SMAW) and Flux Cored Arc Welding (FCAW) [25,26]

3.2 Social Aspects

To consider the social aspects related to the SMAW process, it is important to consider the health hazards and risks involved during this operation. Arc welding operation generates fumes with the airborne particle size of 0.1 – 1.0 μm. The welding fumes mainly consist of heavy metals such as chromium (Cr), manganese (Mn), nickel (Ni) and iron (Fe) [27]. Due to the health related hazards, several cases of eye-nose-throat (ENT) infection, chest congestion, bronchitis, pneumonitis, chronical asthma and lungs cancer [28]. Chang et al. [29] conducted a study where different welding processes such as metal arc welding, manual / automatic gas metal arc welding, laser arc hybrid welding, were investigated for their environmental and social impact using life cycle analysis (LCA) approach. The study highlighted that welder salary is fair, but there are serious concerns towards welders own health during the manual welding operation. Higher health risk was reported for manual metal arc welding process. Welding fumes are associated as one of the major risks towards the health of human worker. Slagor et al. [30] investigated the possibility of cataract in the Danish welders by considering the group of 4288 welders. The study revealed that no significant case of cataract was found in the cohort, that means that by using proper safety equipment this possibility can be eradicated. Hexavalent chromium (CrVl) is also present in the welding fumes and it is main cause of lung diseases such as cancer and occupational asthma. Incase if the CrVl is inhaled from the breathing zone of the welder, it can attach the lungs tissue due to its highly reactive nature [31]. It has also been found that during the welding of carbon steel, Manganese (Mn) is released in the welding fumes [32]. Literature has reported the cases of Parkinson and brain damage due to the exposure of Manganese (Mn) [33].

The welding fumes generation rate is mainly based on the chemical composition of the filler material. The percentage of these welding fumes is reported to be around 90 % to 95 %, generated from the filler metal and the flux coating from the electrode material. The generated fumes rate is dependent on the welding rate, welding process, type of coating material, welding Position, welding environment Chromium content and flux ingredients [34]. The literature also points out at the important precautionary measures to reduce the harmful effects of the welding fumes. As per the OSHA occupational exposure limit (OEL) based guidelines, it is important to regularly take the sample of welding fumes and incorporate the sampling head with filter appropriately in the breathing zone of the welder[35,36]. Similarly, it is very important for the factory floor with welding capabilities to have appropriate ventilation facilities. When it comes to the social aspects of welding operation, literature also points out at the salary of the welder as an important parameter. Several studies have been focused to investigate the salary based social aspects. The study performed in the Germany reveals that welder's salary is sufficient enough to sustain a good quality life [29].

3.3 Environmental Aspects

Due to the generation of hazardous fumes during the welding operation, it is understood that welding operation increases the air pollution. As per the conventional understanding of air pollution, any hazardous particulate or biological molecule that contaminates the natural Earth's atmosphere is considered as an air pollutant. Reduction of welding fumes can decrease the related air pollution. The literature supports the idea of reducing current intensity to lower the emissions from welding fumes [37]. To improve the environmental performance of the SMAW process, one of the key strategies is to lower the energy consumption in the process. Several researchers have investigated this aspect of the SMAW process. Reduction in the energy consumption of SMAW process, bring several environmental and economic benefits to the overall performance. Sterjovski et al. [38] welded 35 mm steel plates and examined the arc energy heat input for the manual arc welding process. The study investigated the sensitivity and reliability of the parameters required to calculate heat input. The heat input is the measure of total power of arc energy produced to perform a welding of unit length. The study also performed Vickers hardness (HV) and optical microscopy on the welded samples. The results from microstructure and hardness evaluation revealed heat input is linked with cooling rate. Sproesser et al. [15] utilized life cycle analysis (LCA) to compare the environmental impact of different welding processes such as manual metal arc welding (MMAW), gas metal arc welding (GMAW), laser arc hybrid welding (LAHW), spray and modified arc welding processes etc. The study incorporated welding operations on 20 mm thick structural steel with the weld bead length of 1 m. The LCA study revealed highest environmental impact for the MMAW process as shown in the Figure 4. The advent of modern power electronics has enabled industry to provide more output power as compared to the conventional transformer based sources. For example, inverter based technology can be utilized to replace the old conventional power sources. The modern technology can deliver more efficient and smoother supply of output power [22].

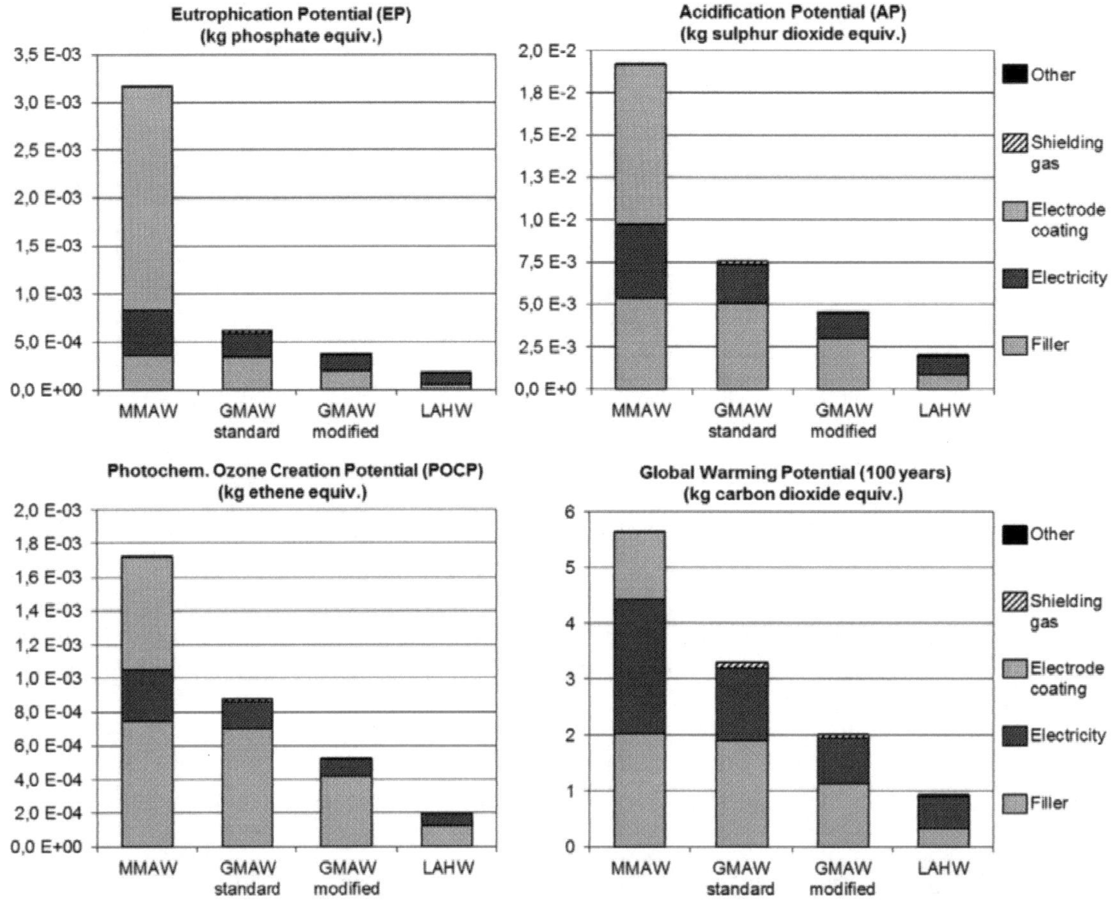

Figure 4. Comparison of environmental impacts for different welding operations [15]

4. Future recommendations and conclusions

The current study characterizes sustainability involved in the SMAW process by using triple bottom line or three pillars of sustainability approach. The study incorporates the available literature to establish the input parameters and output streams involved in the SMAW process. As a result of this current study, several conclusions can be drawn.

- It was found that 80 – 85% of the overall cost in welding operation is related to the labor and other over heads. To maximize the economic gain, different optimization methods can be employed to reduce the resource consumption and cost involved in the SMAW process.
- Fume inhalation by the welder is one of the major health hazards present in the SMAW operation. The risks of this hazard can be significantly reduced by incorporating proper designing and planning of the SMAW process.

- The advent of modern power electronics has permitted industry to provide more output power as compared to the conventional transformer based sources. It means that by using modern power electronics there is a huge potential present for the reduction in energy consumption.
- It has been observed that there are very limited studies available in the literature where life cycle analysis (LCA) of a welding operation has been performed. Further studies can be performed to implement LCA for welding operations.
- The study observed that very limited literature is available in the area of numerical modeling of a welding operations. As future recommendation, more studies can be performed to develop more precise numerical models based on the coupling of Multiphysics such as heat transfer, fluid flow and electromagnetics.
- As the welding process involves solidification and melting of metal, it is important to capture the proper physic and phase change in the numerical model to attain accurate performance predictions.

5. References

[1] Khan I 2009 *Welding Science and Technology* (New Age International Pvt Ltd Publishers)
[2] DuPont J N, Babu S and Liu S 2013 Welding of Materials for Energy Applications *Metall. Mater. Trans. A* **44** 3385–410
[3] Jawahir I.S. and Dillon Jr. O W 2007 Sustainable Manufacturing Processes: New Challenges for Developing Predictive Models and Optimization Techniques *Proceedings of the 1st International Conference on Sustainable Manufacturing (SM1)* (Montreal, Canada)
[4] Chetan, Ghosh S and Venkateswara Rao P 2015 Application of sustainable techniques in metal cutting for enhanced machinability: a review *J. Clean. Prod.* **100** 17–34
[5] Kaebernick H and Kara S 2006 Environmentally sustainable manufacturing: a survey on industry practices *Proc. 13th CIRP ...* 19–28
[6] Lélé S M 1991 Sustainable development: A critical review *World Dev.* **19** 607–21
[7] Gutowski T G 2011 Manufacturing and the Science of Sustainability 1–8
[8] Thiriez A and Gutowski T 2002 An Environmental Analysis of Injection Molding
[9] Jayal A D, Badurdeen F, Dillon O W and Jawahir I S 2010 Sustainable manufacturing: Modeling and optimization challenges at the product, process and system levels *CIRP J. Manuf. Sci. Technol.* **2** 144–52
[10] Hacking T and Guthrie P 2008 A framework for clarifying the meaning of Triple Bottom-Line , Integrated , and Sustainability Assessment *Environ. Impact Assess. Rev.* **28** 73–89
[11] Jenney C L and O'Brien A 1991 Welding handbook *Am. Weld. Soc.* **1** 982
[12] Weman K 2012 *Welding Processes Handbook*
[13] Groover M P 2008 Automation, Production Systems, and Computer-integrated Manufacturing *Automation, Production Systems, and Computer-integrated Manufacturing* p 290
[14] Majumdar J D 2006 Underwater welding-present status and future scope *J. Nav. Archit. Mar. Eng.* **3** 38–47
[15] Sproesser G, Chang Y J, Pittner A, Finkbeiner M and Rethmeier M 2015 Life Cycle Assessment of welding technologies for thick metal plate welds *J. Clean. Prod.* **108** 46–53
[16] Brundtland G H 1987 World Commission on Environment and Development *Our common Futur.* **42/187** 8–9
[17] Elkington J 2001 Enter the Triple Bottom Line *Triple Bottom Line Does it all Add Up?* **1** 1–16
[18] Norman W and Macdonald C 2004 Getting to the Bottom of "Triple Bottom Line" *Bus. Ethics Q.* **14** 243–62
[19] Slaper T and Hall T 2011 The Triple Bottom Line : What Is It and How Does It Work? *Indiana Bus. Rev.* 4–8
[20] Gimenez C, Sierra V and Rodon J 2012 Sustainable operations: Their impact on the triple bottom line *Int. J. Prod. Econ.* **140** 149–59
[21] Parkin S 2000 Sustainable development: the concept and the practical challenge *Institution of Civil Engineers - Civil Engineering* pp 3–8
[22] Matt Albright Reduce Energy Consumption in Your Welding Operations *Lincoln Electr. Co.*

[23] Vimal K E K, Vinodh S and Raja A 2015 Modelling, assessment and deployment of strategies for ensuring sustainable shielded metal arc welding process - A case study *J. Clean. Prod.* **93** 364–77

[24] Jármai K, Snyman J A and Farkas J 2006 Minimum cost design of a welded orthogonally stiffened cylindrical shell *Comput. Struct.* **84** 787–97

[25] Patrick C W 2014 Understanding welding cost – using flux-cored arc welding (fcaw) for cost reduction and productivity improvement *Proceedings of the ASME 2014 Pressure Vessels & Piping Conference.* (July 20-24, 2014, Anaheim, California, USA) pp 1–8

[26] Newell W F 2011 Flux Cored Wires for High Integrity Applications *Proceedings of the 6th Conference on Advances in Materials Technology for Fossil Power Plants, ASM International.* pp 1030–44

[27] J.M. Antonini, S. Stone, J.R. Roberts, B. Chen, D. Schwegler-Berry, A. Afshari D G F 2007 Effect of short-term stainless steel welding fume inhalation expo- sure on lung inflammation, injury, and defense responses in rats *Toxicol. Appl. Pharmacol.* **223** 234–245

[28] Publication D (NIOSH) 1988 Criteria for a Recommended Standard *National Institute for Occupational Safety and Health* pp 88–110

[29] Chang Y-J, Sproesser G, Neugebauer S, Wolf K, Scheumann R, Pittner A, Rethmeier M and Finkbeiner M 2015 Environmental and Social Life Cycle Assessment of welding technologies *Procedia CIRP* **26** 293–8

[30] Slagor R M, La Cour M and Bonde J P 2016 The risk of cataract in relation to metal arc welding *Scand. J. Work. Environ. Heal.* **42** 447–53

[31] Mike Harris Welding Safety: Protecting Employees from the Dangerous Effects of Overexposure to Welding Fumes

[32] Chandra SV, Shukla GS, Srivastava RS, Singh H G V 1981 An exploratory study of manganese exposure to welders. *Clin Toxicol.* **18** 407–16

[33] Racette, B.A., Tabbal, S. D., Jennings, D., Good, L., Perlmutter, J., S., and Evanoff B 2005 Prevalence of parkinsonism and relationship to exposure in a large sample ofAlabama welders.. *Neurology* **64** 230 – 235

[34] Spear J E 2010 Welding Safety: Assessing and Controlling Exposure to Hexavalent Chromium *ASSE Prof. Saf.* 22–32

[35] Sheet O F Controlling Hazardous Fume and Gases during Welding

[36] Hariri A, Yusof M Z M and Leman a. M 2012 Sampling Method for Welding Fumes and Toxic Gases in Malaysian Small and Medium Enterprises (SMEs) *Energy Environ. Res.* **2** 13–20

[37] H N, JY S and M B 2012 Environmental Impacts of Using Welding Gas *J. Technol. Manag. Appl. Eng.* **28** 1–11

[38] Sterjovski Z, Norrish J, Sloan G, Nolan D J, Sterjovski Z, Norrish J, Holdstock R, Sloan G and Nolan D 2005 Evaluation of calculated arc energy heat input in MMAW processes Evaluation of calculated arc energy heat input in MMAW processes *Australas. Weld. J. - Weld. Res. Suppl.* **50** 42–8

Investigation on burrs in micro milling of stainless steel 310S

YUAN Meixia[1,2,3] LIU Shaonan[1,2,3] Xue Hongxin[1,2,3] Tang Boyan[1,2,3]Zhao Linlin[1]

1 School of Mechanical-electronic and Automobile Engineering, Beijing University of Civil Engineering and Architecture, Beijing 100044
2 Beijing Key Laboratory of Performance Guarantee on Urban Rail Transit Vehicles, Beijing 100044
3 Beijing Engineering Research Center of Monitoring for Construction Safety, Beijing 100044
E-mail: yuanmeixia@bucea.edu.cn

Abstract. In the process of micro milling stainless steel 310S, the influence of milling parameters (cutting depth, a_p, feed per tooth f_z, cutting speed v),and down and up milling method were revealed, which provides reference for controlling burrs of stainless steel 310S, improving surface quality and optimizing cutting process. Based on the orthogonal test method, the coated carbide micro diameter cutter was used and milling experiments were carried out on stainless steel 310S.Top burr size data information was collected and analyzed. Up milling is better than down milling because the shape and size of burr are relatively small. With the increase of cutting depth, the shape of burrs appears long fibrous and tearing and wavy serrated, which means the burr getting large and worse. In order to minimize burrs, it is the good way to choose the sharp cutting tools and up milling, control the cutting depth and select feed rate.

1. Introduction

Burr is a serious defect affecting the surface quality of parts, especially for micro parts. If the burr in the cutting process cannot be effectively controlled, it is easy to cause the parts coordination failure or even dimensional error. Therefore, the formation mechanism and control process of burr are the hotspot and research focus of micro cutting. Many domestic and foreign scholars have done a lot of experiments on burrs [1-8]. Fu[9] considered that in the process of cutting, burr generally refer to tiny, prominent residues on the machined surface or the edges of the workpiece. Luo[10] thought that burr generation was due to the undesirable separation of chips and workpiece. Burr was divided into eight categories according to the location of the burr, the up milling edge burr and side burr were studied emphatically, and both of them have the greatest influence on the workpiece. It is considered that the friction angle of the tool rake face causes serious deformation in the cutting zone, resulting in greater burr, so the sharpness of cutting tools greatly affects the generation of burr. The separation of the chip at the workpiece boundary is not complete, or the boundary contour remaining after separation is not an ideal contour, tiny burr is formed. Gillespie[11] divided burr into three basic models: transverse deformation of workpiece, chip bending and chip breaking. The burr formation mechanism of micro cutting is different from that of conventional machining, due to the influence of the minimum cutting thickness, the negative shear zone results in the edge collapse of the workpiece and the plastic deformation resulting in micro burr. In the experiment of micro milling Ni-base superalloy, Lu[12] analyzed the cause of burr formation, and considered that the tool wear affected the chip breaking performance, and the uncut chip accumulation eventually formed strip burr.

Stainless steel 310S is austenitic chromium nickel stainless steel, with a relatively high percentage of chromium and nickel, so 310S has strong creep strength, good heat resistance, suitable for working continuously under high temperature, the material has good corrosion resistance, oxidation resistance, commonly used in steam turbine, aviation machine parts. Micro blade not only requires high profile accuracy, but also requires high surface accuracy. Micro milling cutting tool is generally used for micro milling. In this paper, the surface burrs of stainless steel 310S have been studied. The test results have reference significance for controlling the burrs of stainless steel 310S and improving its surface quality.

2. Formation mechanism of micro cutting burr

The formation mechanism of burrs in micro milling is considered as the scale effect of micro cutting, which results in the formation of negative shear zone[13], as shown in Figure 1, there are three deformation zones in the cutting process. In micro cutting process, due to the arc radius of cutting edge and cutting thickness are in an order of magnitude, compared with the conventional cutting process, the cutting deformation area will be changed, that is the negative shear IV area. In addition, the stiffness of the end support of the workpiece is relatively low, which will also affect the negative shear area[14]. The β is the negative shear angle, and its size determines the amount of deformation at the end of the workpiece. The angle formed by the negative shear plane OD and the machined plane is the negative shear angle, and its size also affects the shape and size of the final burrs formation. The end deformation increases with the increase of the negative shear angle, and the final thickness and height of the burrs increase accordingly.

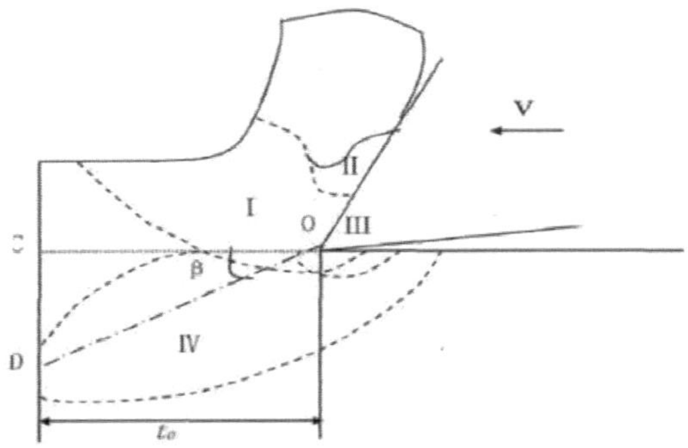

Figure 1. Diagram of generation of negative shear region.

3. Experiment design

The machining center used in this experiment was Carver-45-400. Worktable size: 490×430mm. The cutter is made of Sumitomo carbide with micro diameter, the model is GLM2005SF, the diameter of cutter is ϕ0.5mm, cutting edge length is 1.25mm, shank diameter is 3mm. The workpiece material is stainless steel 310S aluminum alloy.

The factors and levels of the orthogonal test are shown in Table 1.

Table 1. The factors and levels of the orthogonal test.

Factors and levels	A Spindle speed n(r/min)	B Milling depth a_p(mm)	C Feed per tooth f_z (mm/z)
1	6000	0.01	0.001
2	7000	0.03	0.002
3	8000	0.05	0.003
4	9000	0.07	0.004

IOP Conf. Series: Materials Science and Engineering **244** (2017) 012002 doi:10.1088/1757-899X/244/1/012002

KEYENCE microscope system was used to measure and observe the size and shape of the burr. The observed burr measurements are shown in Figure 2. The size and shape of burrs along each groove at different positions are not exactly the same, at the same time, in order to avoid measurement errors, the top burrs of each groove in the eight positions of the up and down milling side were selected respectively. By measuring its width, the average value was then obtained, calculated by equation 1[15].

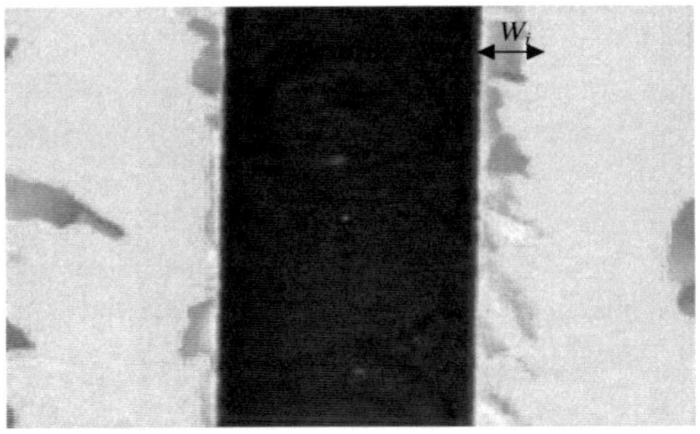

Figure 2. Burr size measurement.

$$W = \frac{\sum_{i=1}^{8} \Delta W_i}{8} \tag{1}$$

Where W represents the average width of the top burr; ΔW_i represents the width of the top burr at each position.

4. Results and analysis of burr test of stainless steel 310S

This experiment was carried out strictly according to orthogonal design of micro milling process. Through data arrangement, measurement and analysis, the results of orthogonal test of top burr size were obtained. The burr width of down and up milling are shown in Table 2. The burr width of up milling is smaller than that of down milling, and the quality of up milling workpiece is higher. Therefore, only the analysis of burr width of up milling is carried out in the subsequent data analysis.

Table 2. Orthogonal test results of slot top burr size

Test number	A Spindle speed n(r/min))	B Milling depth a_p(mm)	C Feed per tooth f_z (mm/z)	Width of burr ω(μm) (up milling)	Width of burr ω(μm)(down milling)
1	6000	0.01	0.001	83.9292	106.198
2	6000	0.03	0.002	116.028	118.485
3	6000	0.05	0.002	107.303	91.9902
4	6000	0.07	0.004	90.3633	83.79475
5	7000	0.01	0.002	237.341	165.826
6	7000	0.03	0.001	155.222	134.522
7	7000	0.05	0.004	76.393	149.151
8	7000	0.07	0.003	77.536	204.433
9	8000	0.01	0.003	136.029	70.655
10	8000	0.03	0.004	116.736	140.360
11	8000	0.05	0.001	99.532	198.509
12	8000	0.07	0.002	77.038	163.639
13	9000	0.01	0.004	138.050	78.363
14	9000	0.03	0.003	70.236	86.806
15	9000	0.05	0.002	114.982	119.652
16	9000	0.07	0.001	137.450	93.292

4.1. Range analysis of measurement results based on orthogonal test method

Range analysis, abbreviation is R method, is also called visual analysis. Through the range analysis, the primary and secondary order of the influence of each factor on the experimental indexes was obtained. The optimal level of each experimental factor, the optimal combination of processing parameters and the experimental indexes vary with each experimental factor. K_{ji} represents the sum of all test indexes value corresponding to the test factors of column j on the level of factor i. $\overline{K_{ji}}$ represents the average value of K_{ji}, By comparing the size of the $\overline{K_{ji}}$, the optimal level of the experimental factors in the group can be determined. The optimal combination of experimental factors was formed by combining the optimal levels of the experimental factors in each group.

$$R_j = \max(\overline{K_{j1}}, \overline{K_{j2}}, \ldots\overline{K_{ji}}) - \min(\overline{K_{j1}}, \overline{K_{j2}}, \ldots\overline{K_{ji}}) \quad (2)$$

R_j reflects the influence of experimental factors on experimental indexes. The greater of the value of R_j, the greater influence of the experimental factors on the experimental indexes, and the smaller of the value of R_j, the opposite.

Table 3. Range analysis of orthogonal test (up milling).

Serial number	A Spindle speed n(r/min)	B Milling depth a_p(mm)	C Feed per tooth f_z（mm/z）
K_{j1}	397.624 (6000)	595.349 (0.01)	476.133 (0.012)
K_{j2}	546.492 (7000)	458.222 (0.03)	545.389 (0.028)
K_{j3}	429.335 (8000)	398.21 (0.05)	391.104 (0.048)
K_{j4}	460.718 (9000)	382.387 (0.07)	421.542 (0.072)
$\overline{K_{j1}}$	99.406	148.84	119.033
$\overline{K_{j2}}$	136.623	114.556	136.347
$\overline{K_{J3}}$	107.334	99.553	97.776
$\overline{K_{j4}}$	115.180	95.597	105.386
R_j	37.217	53.243	38.571
Primary and secondary order		$R_b > R_c > R_a$	
Better level	A_1	B_4	C_3
Optimal combination		$A_1B_4C_3$	

Based on the results of orthogonal tests in Table 2, a range analysis of Table 3 can be obtained. It can be seen from the range analysis that the magnitude of the range R_j is $R_b > R_c > R_a$. The results show that the main order of influence of each experimental factor on the burr size of micro milling groove is milling depth > feed rate > spindle speed. Therefore, within the range of the experimental level, the influence of cutting depth on the surface roughness of the test index is the biggest, followed by the feed rate, and the speed has the least influence on the burr size. By the $\overline{K_{ji}}$ of each experimental factor, it can be seen that the optimum machining parameter combination is A1B4C3. It means that when the spindle speed is 9000r/min, the milling depth is 0.01 mm and the feed per tooth is 0.072mm/z, the up milling burr size of the micro milling can be better controlled.

In order to make a more intuitive analysis of the relationship between the experimental factors and the experimental indexes, according to Table 2 and Table 3, the influence trend of the cutting parameters on the burr size at the top of the groove is obtained. The abscissa in the figure is different values of each experimental factor. The ordinate is the top burr size of the workpiece ω, and the smaller of the ω value, the smaller of the burr size; otherwise, the larger of the burr size.

4.2. Influence of spindle speed on burr size of micro groove

In the up milling process, with the increase of spindle speed, the top burr size of workpiece groove firstly increases, then decreases and then increases, and the overall trend is increasing. As shown in Figure 3, when the spindle speed reaches 7000r/min, the top burr size is the maximum. When the spindle speed continues to increase, the top burr size tends to decrease, and the size of top burr increases with the further increase of the speed. The reason is with the increase of spindle speed, the increase of vibration of up milling spindle leads to lower cutting stability, cutting is not uniform; at the same time, because the wear of cutting tool is aggravated after processing, the top burr size is increased as well.

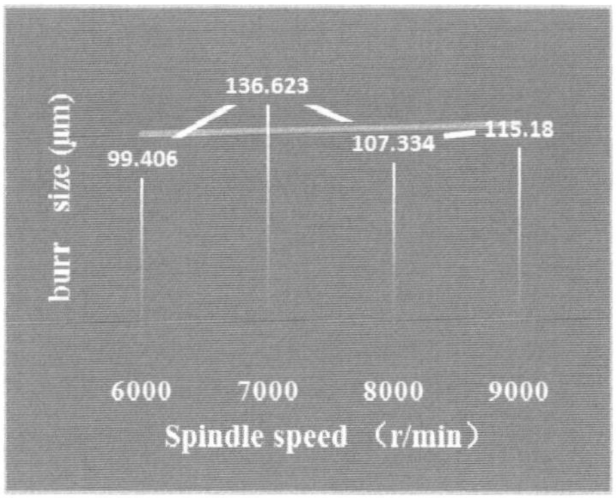

Figure 3. The influence of the spindle speed at the up milling side on the top burr size of the micro groove.

4.3. Influence of spindle speed on burr size of micro groove

As shown in Figure 4, the top burr size of the micro groove in the up milling side decreases as a whole. When the milling depth reaches 0.01mm, the burr size will be the maximum. With the increase of cutting depth, the size effect decreases, forming continuous chips, and the top burr size of the micro groove decreases.

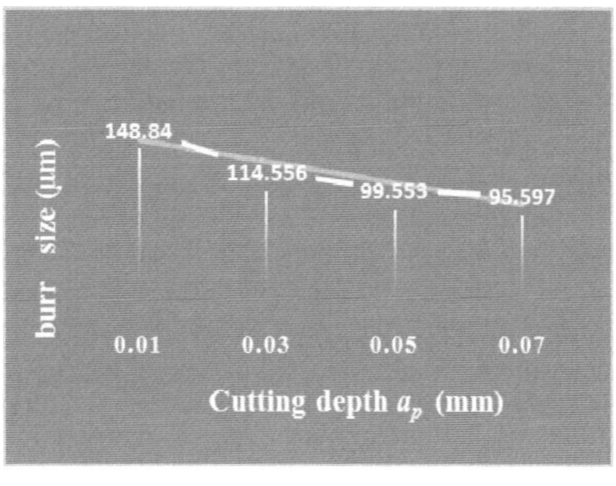

Figure 4. The influence of the cutting depth at the up milling side on the top burr size of the micro groove.

4.4. Influence of feed per tooth on burr size of micro groove

In the up milling side, when the feed rate of each tooth increases continuously, the top burr size of the micro groove first increases and then decreases. It can be seen from the Figure 5 that when the feed rate is small, the size of the top burr increases as the feed per tooth increases, when the feed per tooth is 0.028mm/z, the size of top burr reaches the maximum 136.347mm. With the further increase of feed per tooth, the top burr size decreased, and when the feed per tooth is 0.048mm/z, the top burr size is the lowest. With the further increase of feed per tooth, the cutting thickness increases, the machined surface residual area increases, and the cutter cutting resistance increases, which increases the plastic deformation of the workpiece and leads to the increase of burr size.

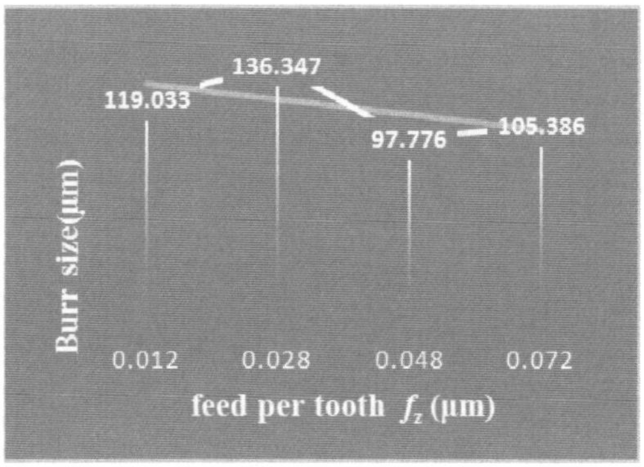

Figure 5. The influence of the feed per tooth at the up milling side on the top burr size of the micro groove.

5. Conclusion

Taking the top burr size of micro milling groove as the experimental index, the micro milling experiments of stainless steel 310S were carried out with the spindle speed, the milling depth and the feed amount as the experimental factors. Through range analysis, the following conclusions are drawn:

（1）Up milling is better than down milling because the shape and size of burr are relatively small.

（2）The main order of influence of cutting parameters on the burr size of micro milling groove is milling depth > feed rate > spindle speed. Within the range of the selected experimental level, the milling depth has the biggest influence on the top burr size of the micro milling groove, followed by the feed per tooth, and the spindle speed has the least influence on the burr size of the up milling side of the micro milling groove.

（3）The effect of cutting depth on the burr size of micro groove milling is very important. With the increase of cutting depth, shape and size of the burr change greatly, converted from a long strip to a wavy zigzag, So the range of cutting depth should be controlled as much as possible.

Fund: Suported by Beijing Municipal Organization Department (Item no. 2014000020124G056);Beijing Municipal Commission of Education (Item no. KM201510016008)

6. References

[1] T Zhang, Z Q Liu, C H Xu. Influence of Size Effect on Burr Formation in Micro Cutting[J]. *Int JAdv Manuf Technol*, 2013, **170**:4801-08

[2] D. Biermann, M. Steiner. Analysis of Micro Burr Formation in Austenitic Stainless Steel X5CrNi18-10[J]. *Procedia CIRP*, 2012, **3**:97-102

[3] ZHANG Haijun , HUANG Yanhua , YUAN Guanghui , TAO Yang , LIU Feng. Experimental Study on Burrs in Micro Milling ICF Micro Targets. *Technology and test*, 2012(**07**)

[4] Tao Zhang, Zhanqiang Liu, Chonghai Xu. Influence of Size Effect on Burr Formation in Micro Cutting[J]. *The International Journal of Advanced Manufacturing Technology*, 2013 (9-12)
[5] Seyed Ali Niknam, Victor Songmene. Simultaneous Optimization of Burrs Size and Surface Finish When Milling 6061-T6 Aluminium Alloy[J] . *International Journal of Precision Engineering and Manufacturing.*, 2013 (**8**)
[6] L.Zhou, Y.Wang, Z.Y.Ma, X.L. Yu. Finite Element and Experimental Studies of the Formation Mechanism of Edge Defects During Machining of SiCp/Al Composites[J]. *International Journal of Machine Tools and Manufacture*, 2014, 84 (**6**): 9-16
[7] Wen Jun Deng, Zi Chun Xie, Ping Lin, Tong Kui Xu, Ming-Xing Zhang. Study on Burr Formation at the Top Edge in Rectangular Groove Cutting[J]. *Advances in Materials Science and Engineering*, 2012
[8] Seyed Ali Niknam, Victor Songmene. Analytical Modelling of Groove Milling Exit Burr Size[J]. *The International Journal of Advanced Manufacturing Technology*, 2014 (1-4)
[9] FU Tian-jiao. Research on Formation and Control Technology of Micro - milling Burr[D].*Shan Dong University*, 2014
[10] Luo Meng. Mechanism and Control Methods of Burr Formation in Metal Cutting Process[D]. *Shang Hai: Shang Hai Jiao Tong University*, 2007
[11] L.K.Gilespie, P.T.Blotter.The Formation and Properties of Machining Burrs[J]. *ASME J.Eng.Ind*, 1976, 98(**1**):66-74
[12] LU Xiaohong, Wang Wenyi, Wang Wentao, et al. Experimental Study on Burrs in Micro-milling Nickel-base Superalloy Inconel718[J]. *Modular Machine Tool & Automatic Manufacturing Technology*, 2015(**1**):1-3
[13] YANG Kai, BAI Qingshun, Yu Fuli, Modelling and Experimental Analysis of the Mechanism of Micro-burr Formation and Micro-end-milling Process[J]. *Nanotechnology and Precision Engineering*, 2010, 8(**1**): 75-83
[14] Zhu Yunming, Wang Jingui, Wang Guicheng, et.al. Study on Simulation of Turning Burr Formation Based on Finite Element Method[J]. *Journal of System Simulation*, 2015, 27(**5**):1120-1126
[15] NI Hai-bo. Simulation and Experimental Study on the Forming Process of Micro-milling Burr[D]. *Harbin:Harbin Institute of Technology*, 2012

IOP Conf. Series: Materials Science and Engineering **244** (2017) 012003 doi:10.1088/1757-899X/244/1/012003

Prediction of multi performance characteristics of wire EDM process using grey ANFIS

Somasundaram Kumanan[1] and Anish Nair[1]
1 Production Engineering Department, National Institute of Technology Tiruchirappalli, 620015, India

Abstract. Super alloys are used to fabricate components in ultra-supercritical power plants. These hard to machine materials are processed using non-traditional machining methods like Wire cut electrical discharge machining and needs attention. This paper details about multi performance optimization of wire EDM process using Grey ANFIS. Experiments are designed to establish the performance characteristics of wire EDM such as surface roughness, material removal rate, wire wear rate and geometric tolerances. The control parameters are pulse on time, pulse off time, current, voltage, flushing pressure, wire tension, table feed and wire speed. Grey relational analysis is employed to optimise the multi objectives. Analysis of variance of the grey grades is used to identify the critical parameters. A regression model is developed and used to generate datasets for the training of proposed adaptive neuro fuzzy inference system. The developed prediction model is tested for its prediction ability.

1. Introduction

Newer materials are continuously being developed to cater to the needs of ultra-supercritical power plants to withstand temperature up to 975K and pressure up to 26MPa. Some of these materials are Inconel 617, Nimonic 263, Inconel 740, Haynes 230, Haynes 282. Inconel 617 is the lead candidate material(1) falls into the Group D category of Nickel base Superalloys. The conventional machining of superalloys are limited due to high chemical reactivity, high hot hardness, rapid strain hardening, segmented chip formation, self-induced chatter, low thermal conductivity and poor heat dissipation at tool chip interface. Poor tool life, accelerated abrasion, high adhesion and higher diffusion wear are the productivity problems faced in the conventional machining of superalloys. The machining of nickel base super alloys needs attention as the wastage while machining becomes unaffordable(2). Non-conventional machining techniques such as EDM and wire EDM are effective alternatives. It has been reported that the first geometries can be effectively machined using wire EDM process(3). Wire EDM process has been suggested as a replacement for broaching for the development of fir tree slots in turbine blades(4). EDM process is the most competitive technology for turbine blisk finishing operation (5). In a product the dimensional deviation, form tolerances and orientation tolerances play a significant role(6). The effect of feed rate, current and pulse on time on the circularity deviation of a hot work tool steel was studied and regression analysis was applied to develop predictive models(7). Electrical discharge machining of conductive ceramics was investigated with focus on form and orientation tolerances using grey relational analysis(8). Grey Relational Analysis (GRA) (9) has been successfully employed to solve problems with high uncertainty, multi-inputs and discrete data. Optimization based on the grey relational technique is one of the commonly adopted methods in machining processes(10). Multiple regression Analysis (MRA) are used to get a second order non-linear equation and these equations are used to create a large dataset for the training and testing of the neuro fuzzy models. A neural network with fuzzy decision rules forms a hybrid intelligent system. ANFIS has been used by researchers to create prediction models for different welding(11) and machining processes (12). Prediction of white layer thickness and surface roughness using ANFIS model were reported in literature (13). The response features based on the form and orientation tolerances needs attention. This paper proposes the Multi performance optimization in wire EDM using grey ANFIS.

2. Proposed methodology

The proposed methodology of development of Grey ANFIS for predicting parameters with multi performance characteristic of wire EDM process is depicted in Figure 1. Parameters under consideration are pulse on time, T_{on} (μs), pulse off time, T_{off} (μs), discharge current (Ampere), peak voltage (Volts), flushing pressure (lpm), wire tension (kgf), table feed (mm/min) and wire speed (mm/min) which lead to an Taguchi L18 experimental design The performance characteristics are surface roughness (μm), material removal rate (mm^3/sec), Wire Wear Rate (gram/sec), and form and orientation tolerances. Accuracy measures such as MAD, MAD and RMSE are used to make the proposed neuro fuzzy model anetwork as a robust prediction model.

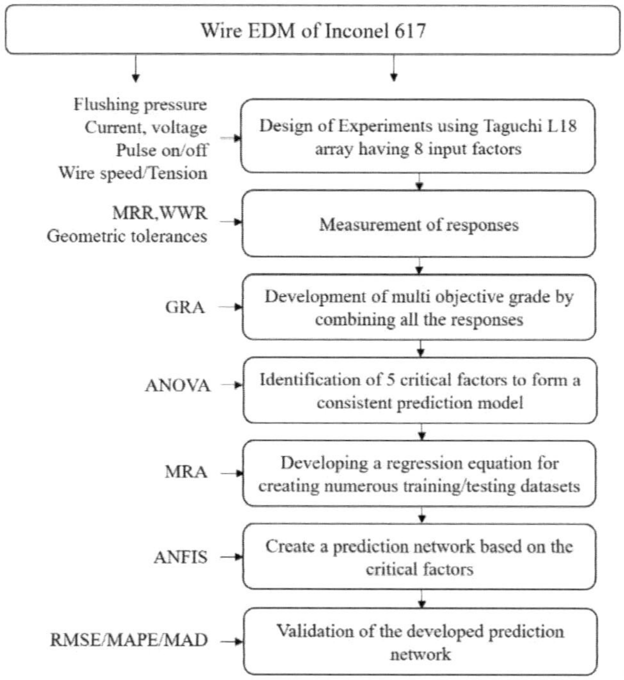

Figure 1. Methodology for the development of a neuro fuzzy prediction model for wire EDM.

3. Experimentation

The proposed experiments are carried out in wire cut EDM machine 'Electronica Sprintcut'. The wire is brass alloy of 0.25mm diameter and the dielectric is deionized water. A semicircular profile comprising of two parallel edges is cut into the 10mm thick Inconel 617 workpiece. The wire EDM machining set up is shown in the Figure 2. The Taguchi L_{18} mixed level orthogonal array is planned as per the principles of design of experiments. Flushing pressure has been set at two levels and the rest of the parameters are set at three levels each as show in the Table 1. The experiments are conducted and the various results obtained are shown in Table 2.

Figure 2. Machining of the profile in Electronica Sprintcut.

Table 1. Experimental design-L18 array for wire EDM of Inconel 617.

S.No.	Input Parameters							
	Flushing (lpm)	Table feed (mm/min)	T_{on} (μs)	T_{off} (μs)	Current (A)	Voltage (Volts)	Tension (kgf)	Wire Speed (mm/min)
1	6	2.70	110	50	150	10	3	3
2	6	2.70	125	55	200	15	8	4
3	6	2.70	131	63	230	20	14	5
4	6	3.83	110	50	200	15	14	5
5	6	3.83	125	55	230	20	3	3
6	6	3.83	131	63	150	10	8	4
7	6	5.18	110	55	150	20	8	5
8	6	5.18	125	63	200	10	14	3
9	6	5.18	131	50	230	15	3	4
10	12	2.70	110	63	230	15	8	3
11	12	2.70	125	50	150	20	14	4
12	12	2.70	131	55	200	10	3	5
13	12	3.83	110	55	230	10	14	4
14	12	3.83	125	63	150	15	3	5
15	12	3.83	131	50	200	20	8	3
16	12	5.18	110	63	200	20	3	4
17	12	5.18	125	50	230	10	8	5
18	12	5.18	131	55	150	15	14	3

Material removal rate is estimated based on the kerf width, which is an outcome of the machining. The kerf width is measured using an Olympus make optical microscope. MRR is obtained from the calculated kerf width values and the cutting speed which is calculated during the machining process as shown in Eq.(1) (14),

$$MRR = KhCS \tag{1}$$

K is the Kerf width, h is the thickness of the workpiece, CS is the cutting speed. The wire wear rate is calculated by the weight loss method. The difference in weight between the fresh wire and the used wire is found out. The rate at which the wire erodes can be calculated from the machining time recorded for each cycle of measurement. The wire wear rate is a direct index of the productivity of the process. The wire images are obtained using the Olympus optical microscope. Surface roughness (R_a) values are measured using the Talysurf Surfcorder probe type tester. R_a values indicate the peak and valleys in a particular line where the probe moves. The surface roughness measurements are made with a cut off value of 0.08mm and are made against the direction in which the wire travels and hence leading to higher values of surface roughness. The circularity error, perpendicularity error, cylindricity error and parallelism error are measured using the HELMEL make coordinate measuring machine. These errors are a direct index of the dimensional deviation which occur during the electrical discharge machining process.

Table 2. Measured responses in the wire EDM of Inconel 617.

No.	MRR	WWR	Ra	Circularity	Cylindricity	Perpendicularity	Parallelism
1	3.7812	0.0150	5.5290	0.0069	0.0120	0.0676	0.0070
2	10.1033	0.1252	7.9310	0.0068	0.0127	0.0955	0.0050
3	10.9371	0.2044	8.6320	0.0154	0.0109	0.0738	0.0096
4	4.7666	0.0306	5.7550	0.0159	0.0134	0.0758	0.0068
5	10.5225	0.1192	8.6670	0.0121	0.0124	0.0837	0.0016
6	8.6970	0.0991	9.0570	0.0099	0.0157	0.0557	0.0062
7	3.1506	0.0206	5.9540	0.0118	0.0177	0.0706	0.0045
8	9.2167	0.1419	8.9430	0.0102	0.0117	0.0905	0.0092
9	12.7500	0.1216	9.6880	0.0127	0.0099	0.0940	0.0106
10	2.3563	0.0082	6.6980	0.0318	0.0140	0.0784	0.0156
11	8.7055	0.1081	7.5160	0.0126	0.0154	0.0841	0.0100
12	9.9794	0.1070	10.3300	0.0204	0.0140	0.0730	0.0097
13	6.5413	0.0604	8.2290	0.0127	0.0110	0.0647	0.0084
14	7.1534	0.0659	7.3570	0.0235	0.0188	0.0533	0.0134
15	14.5051	0.2964	10.7650	0.0115	0.0159	0.0619	0.0101
16	2.2694	0.0027	7.3130	0.0246	0.0178	0.0377	0.0253
17	14.7037	0.2327	10.4760	0.0083	0.0116	0.0630	0.0138
18	11.4241	0.2401	9.3520	0.0170	0.0177	0.0569	0.0195

4. Adaptive neuro fuzzy inference system

4.1. Grey grade development

The normalization of the responses is carried out in order to minimize the variability in the data range. After the normalization process the grey relational coefficients are calculated using the Eq. (4) where Δ_{min} is the smallest value of $\Delta_{0i}(k)$, Δ_{max} is the largest value of $\Delta_{0i}(k)$. The absolute difference between $x_o^*(k)$ and $x_i^*(k)$ is $\Delta_{0i}(k)$. $x_o^*(k)$ is the ideal or reference sequence.

The grey relational coefficient is given by the following formulae,

$$x_i = \left(x_i(1), x_i(2), ..., x_i(n)\right), x_i \in X, i = 1, 2, ..., m \tag{2}$$

x_i be the compared series and x_0 be the reference series

$$x_0 = \left(x_0(1), x_0(2), ..., x_0(n)\right), x_0 \in X \tag{3}$$

The grey relational coefficient is given by,

$$\gamma\left(x_0(k), x_i(k)\right) = \frac{\Delta_{min} + \zeta\Delta_{max}}{\Delta_{0i}(k) + \zeta\Delta_{max}} \tag{4}$$

The grey relational grade is obtained from the following formula,

$$\gamma\left(x_0, x_i\right) = \frac{1}{n}\sum_{k=1}^{n}\gamma\left(x_0(k), x_i(k)\right) \tag{5}$$

The grey relational coefficients are combined into a grey relation grade by giving equal weightage to all the responses. Eq. (5) is used for calculating the grey relational grade. The developed grades are shown in Table 3.

4.2. Analysis of Variance

Analysis of variance of the grey grade is carried out with respect to the eight factors and the results are as tabulated in Table 4. Flushing speed has the highest contribution on the responses followed by pulse on time and pulse off time.

4.3. Multiple regression analysis

The accuracy of a neuro fuzzy model depends on the number of data from which it is created. Since the experimental values here are limited MRA is used for obtaining the training and testing datasets. Using MRA a second order non-linear regression model is developed which relates the grey grades with the five input factors. This model is further used to develop a huge dataset which is further given as input to the neuro fuzzy model. Figure 5 shows the normal residual plot of the regression equation and from the plot it can be seen that deviation from the normal line is uniform indicating a good fit for the values.

$$\text{Regression equation for grey grade} = 2.19 - 0.01528\times A + 0.0082\times C - 0.0296\times D - 0.1372\times F + 0.0399\times G - 0.000056\times C^2 + 0.000270\times D^2 + 0.001330\times F^2 - 0.000897\times G^2 - 0.000083\times C\times D + 0.000477\times C\times F + 0.000116\times C\times G + 0.000691\times D\times F - 0.000628\times D\times G - 0.000414\times F\times G$$

4.4. Neuro fuzzy inference system

Inputs considered for modelling the ANFIS structure are flushing speed, T_{on}, T_{off}, voltage and tension of the wire. The developed grey grade is the output for the ANFIS model. The membership functions are assigned by dividing the input space into fuzzy regions. In this present investigation bell membership function is used. The training data set is used to find the initial premise parameters for the membership functions equally spacing each of the membership functions. The fuzzy rules are generated based on the input and output data pairs attained by experimentation. The conflicting rules are resolved by assigning a degree to each rule. After the conflicting rules are resolved a combined rule base is formed. The final stage is defuzzification process to get the output value. The centroid method is used for defuzzification. The ANFIS structure shown in Figure 4 consists of five layers. The nodes in the layers 1 and 4 are adaptive nodes. Nodes in layers 2, 3 and 5 are fixed nodes

Table 3. Obtained grey grade from the combination of responses.

No.	Flushing speed (lpm)	Table Feed (mm/min)	T_{on} (μs)	T_{off} (μs)	Current (A)	Voltage (Volts)	Tension (kgf)	Wire Speed (mm/min)	Grey Grade
1	0.3627	0.9227	1.0000	0.9921	0.6794	0.4915	0.6870	0.3627	0.7344
2	0.5747	0.5452	0.5215	1.0000	0.6138	0.3333	0.7770	0.5747	0.6243
3	0.6227	0.4213	0.4576	0.5924	0.8165	0.4446	0.5970	0.6227	0.5652
4	0.3849	0.8403	0.9205	0.5787	0.5597	0.4313	0.6950	0.3849	0.6307
5	0.5979	0.5576	0.4548	0.7022	0.6403	0.3858	1.0000	0.5979	0.6204
6	0.5086	0.6037	0.4260	0.8013	0.4341	0.6162	0.7204	0.5086	0.5878
7	0.3499	0.8914	0.8603	0.7143	0.3633	0.4676	0.8034	0.3499	0.6364
8	0.5312	0.5134	0.4340	0.7862	0.7120	0.3537	0.6093	0.5312	0.5634
9	0.7609	0.5526	0.3863	0.6793	1.0000	0.3392	0.5683	0.7609	0.6130
10	0.3349	0.9639	0.6913	0.3333	0.5205	0.4152	0.4584	0.3349	0.5316
11	0.5090	0.5822	0.5685	0.6831	0.4472	0.3838	0.5852	0.5090	0.5375
12	0.5682	0.5847	0.3529	0.4789	0.5205	0.4502	0.5940	0.5682	0.5076
13	0.4324	0.7179	0.4923	0.6793	0.8018	0.5170	0.6354	0.4324	0.6115
14	0.4516	0.6991	0.5888	0.4281	0.3333	0.6494	0.5011	0.4516	0.5222
15	0.9690	0.3333	0.3333	0.7267	0.4258	0.5443	0.5823	0.9690	0.5598
16	0.3333	1.0000	0.5947	0.4125	0.3603	1.0000	0.3333	0.3333	0.5769
17	1.0000	0.3897	0.3461	0.8929	0.7236	0.5332	0.4927	1.0000	0.6261
18	0.6547	0.3822	0.4065	0.5507	0.3633	0.6008	0.3983	0.6547	0.4800

IOP Conf. Series: Materials Science and Engineering **244** (2017) 012003 doi:10.1088/1757-899X/244/1/012003

Table 4. Analysis of variance.

Source		DF	SS	MS	F-Value	% contribution
Flushing (lpm)	A	1	0.0215	0.0215	8.7900	52.375
Table feed (mm/min)	B	2	0.0001	0.0001	0.0300	0.161
T_{on} (μs)	C	2	0.0139	0.0070	2.8500	16.968
T_{off} (μs)	D	2	0.0107	0.0053	2.1800	13.006
Current (A)	E	2	0.0010	0.0005	0.1900	1.158
Voltage (Volts)	F	2	0.0044	0.0022	0.9000	5.367
Tension (kgf)	G	2	0.0037	0.0018	0.7500	4.483
Wire Speed (mm/min)	H	2	0.0004	0.0002	0.0900	0.523

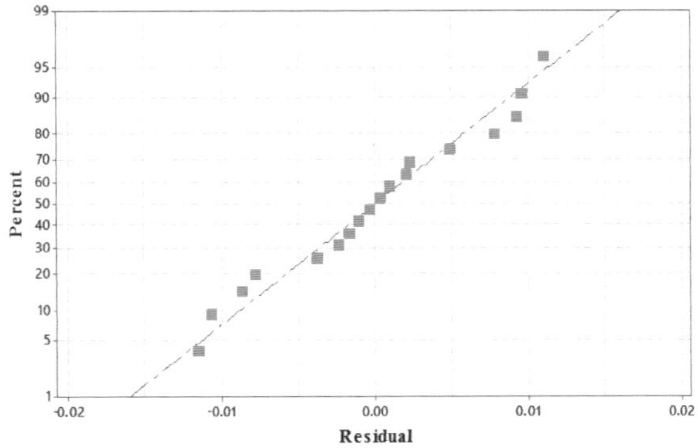

Figure 3. Normal residual plot for the obtained regression equation.

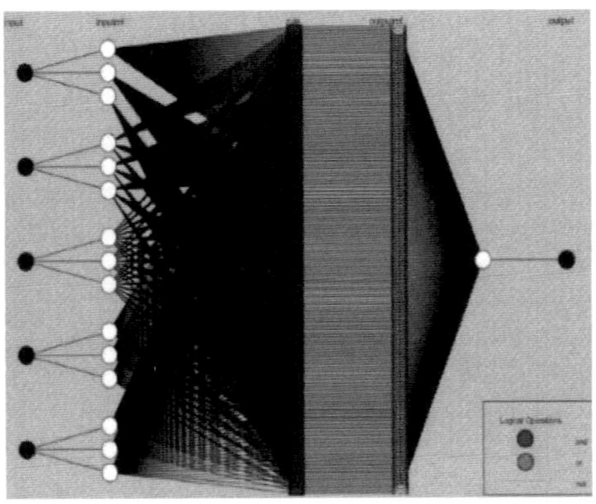

Figure 4. Neuro fuzzy structure for the development of prediction model.

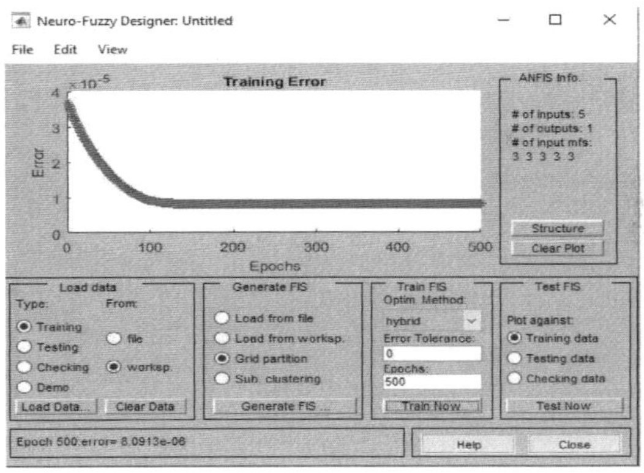

Figure 5. Typical ANFIS editor for the multi performance analysis.

The model is developed with five input neurons and a single output neuron with respective membership functions of Sugeno type FIS. The developed ANFIS model with five inputs and a single output parameter is trained and training details are presented. The ANFIS architecture employed with 'gbellmf' membership function consists of 243 rules which are generated based on the given input data set. Figure 5 shows the ANFIS editor for the wire EDM of Inconel 617. Figure 6 shows the ANFIS rule viewer with various membership functions used for this investigation. Figure 7 shows the effect of pulse on and pulse off time on the grey grades. Figure 8 shows the effect of pulse on time and flushing pressure on grey grade. It is conspicuous from the illustration that grey grade increases as the pulse on time increases whereas grey grade reduces at lower flushing pressures. Figure 9 shows the effect of pulse off time and flushing pressure on grey grades. It is clearly illustrated that a combination of higher pulse off time and lower flushing pressure leads to maximum grey grade.

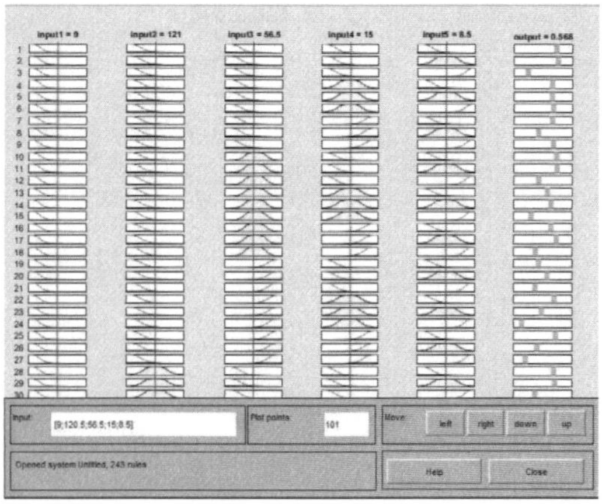

Figure 6. ANFIS rule viewer depicting the rules created.

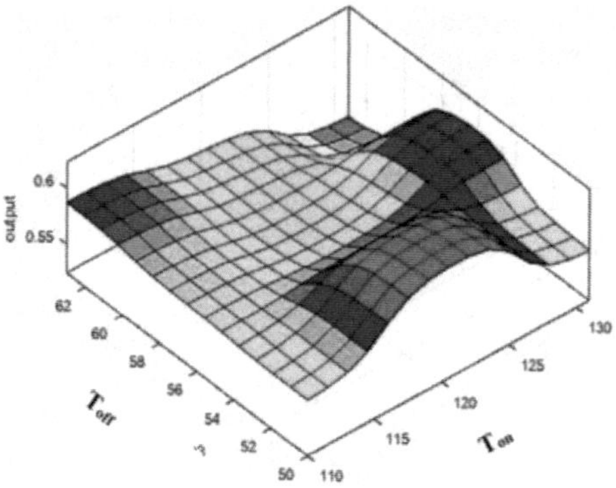

Figure 7. Influence of T_{off} and T_{on} on the grey grade.

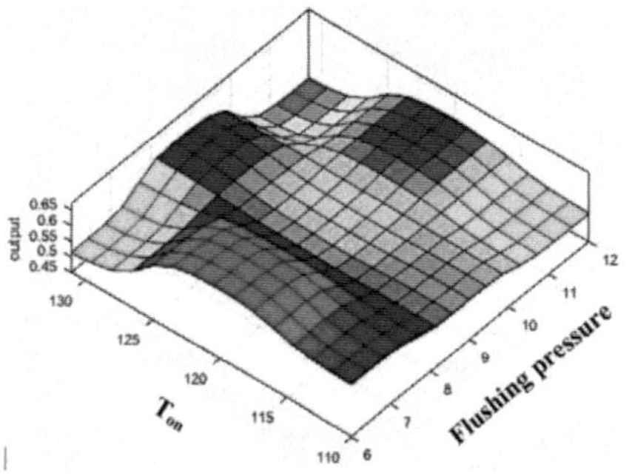

Figure 8. Influence of T_{on} and Flushing pressure on the grey grade.

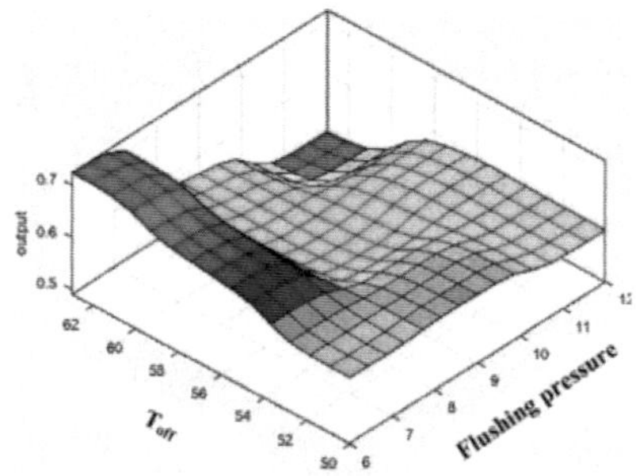

Figure 9. Influence of T_{off} and flushing pressure on the grey grade.

4.5. Performance analysis of the developed neuro fuzzy model

After training the model, it is validated and the results are presented as shown in Figure 10. The values predicted by the developed model are compared with the experimental values and it is revealed that there is a close relationship among the experimental values and the predicted values which are from the developed ANFIS model.

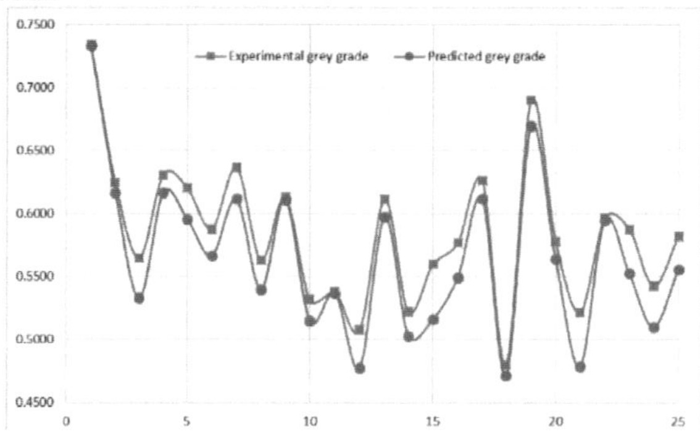

Figure 10. Comparison of the experimental and the predicted grey grades.

The absolute percentage error obtained between the predicted values and the calculated values of Grey relational grade value shows that the developed model is accurate for prediction of desired performance characteristics. The mean absolute deviation value was obtained as 0.0152. Root mean square error is 0.0216 and the Mean absolute percentage error is 2.91. Hence it can be observed that a robust prediction model has been obtained.

5. Conclusion

Limited research has been reported in the wire EDM of Inconel 617 which is a candidate material for ultra-supercritical boiler components. Form and orientation tolerances such as circularity deviation, cylindricity deviation have not been studied extensively and needs attention. The paper investigated the wire EDM process for establishing the control parameters in order to develop an artificial neuro fuzzy Inference model for prediction. Parameters under consideration are pulse on time, T_{on} (μs), pulse off time, T_{off} (μs), discharge current (Ampere), peak voltage (Volts), flushing pressure (lpm), wire tension (kgf), table feed (mm/min) and wire speed (mm/min) which lead to an Taguchi L18 experimental design The performance characteristics are surface roughness (μm), material removal rate (mm^3/sec), Wire Wear Rate (gram/sec), and form and orientation tolerances. The model was successfully developed such that the changes in the input parameter reflects the output. Accuracy measures such as Mean Absolute Deviation, Root Mean Square Error and Mean Absolute Percentage Error were calculated for the neuro fuzzy network indicating a robust prediction model. The mean absolute deviation value was obtained as 0.0152. Root mean square error is 0.0216 and the Mean absolute percentage error is 2.91. Further studies can be carried out in development of multi objective prediction models to reduce the dependence on grey techniques and the results can be compared.

6. References

[1] Narula R. Impacts of steam conditions on plant materials and operations in ultra-supercritical coal power plants. In: D.Zhang, editor. Ultra-Supercritical coal power plants Materials Technologies and optimization. Woodhead Publishing Limited, USA; 2013. p. 23–56.

[2] Wheeldon JM, Shingledecker JP. Materials for boilers operating under supercritical steam conditions. In: Zhang D, editor. Ultra-Supercritical coal power plants Materials Technologies and optimization. Woodhead Publishing, USA; 2013. p. 81–103.

[3] Klocke F, Zeis M, Klink a., Veselovac D. Technological and economical comparison of roughing strategies via milling, sinking-EDM, wire-EDM and ECM for titanium- and nickel-based blisks. CIRP J Manuf Sci Technol 2013;6(3):198–203.

[4] Klocke F, Klink A, Veselovac D, Aspinwall DK, Soo SL, Schmidt M, et al. Turbomachinery component manufacture by application of electrochemical, electro-physical and photonic processes. CIRP Ann - Manuf Technol 2014 63(2):703–26.

[5] Zhan H, Zhao W, Wang G. Manufacturing turbine blisks. Aircr Eng Aerosp Technol. 2011;72(3):247–52.

[6] Manikandan.N, Kumanan S & SC. Optimization of Electrochemical drilling process using Taguchi method and regression analysis. Int J Mach Mach Mater (in Press).

[7] Ugur Esme, M. Kemal Kulekci, Adnan Akkurt US. Mathematical Modeling of Diameter and Circularity Deviation in Wire Electrical Discharge Machining of a Hot Work Tool Steel. Mater Test 2013;55(6):472–7.

[8] Selvarajan L, Narayanan CS, Jeyapaul R. Optimization of Process Parameters to Improve Form and Orientation Tolerances in EDM of MoSi 2 -SiC Composites. Mater Manuf Process 2014;30(8):954–60.

[9] Liu S, Lin Y. Grey Systems: Theory and Applications Berlin, Heidelberg: Springer Berlin Heidelberg; 2011. 1-18 p.

[10] Sreenivasulu R, SrinivasaRao DC. Application of grey relational analysis for surface roughness and roundness error in drilling of Al 6061 alloy. Int J Lean Think. 2012;3(2):67–78.

[11] Raja Dhas JE, Kumanan S. ANFIS for prediction of weld bead width in a submerged arc welding process. J Sci Ind Res (India). 2007;66(4):335–8.

[12] Svalina I, Simunovic G, Simunovic K. Machined Surface Roughness Prediction Using Adaptive Neurofuzzy Inference System. Appl Artif Intell 2013;27(9):803–17.

[13] Çaydaş U, Hasçalik A, Ekici S. An adaptive neuro-fuzzy inference system (ANFIS) model for wire-EDM. Expert Syst Appl. 2009;36(3 PART 2):6135–9.

Study of wear performance of deep drawing tooling

Vishal G Naranje[1], Ram Karthikeyan[2] and Vipin Nair[3]
1,2,3 Birla Institute of Technology, Pilani, Department of Mechanical Engineering Dubai Campus 345055 UAE

Abstract. One of the most common challenges for many of the mechanical engineers and also in the field of materials science is the issue of occurrences of wear of the material parts. In this paper, wear behaviour of particular grade High Carbon High Chromium Steel and many most famously D2, H13, O1 known as theViking steel has been studied, evaluated and analyzed on pin-on-disc machine. The specimens have been subjected under different processing parameters such as speed, load, track diameter and time under dry sliding conditions at normal room temperature to replicate the deep drawing process. The present research work will help to investigate the wear behaviour of the deep drawing die tool to characterize the mechanical & tribological properties of die steel and developed a statistical model to predict wear under dry conditions for deep drawing process.

1. Introduction

Deep drawing is one of the typical sheet metal forming process used to produce axisymmetric and irregular hollow shaped parts through combined action of tensile and compressive deformation of sheet metal blank. It is popular because of its rapid press cycle to produce complex axisymmetric geometries [1]. The life of deep drawing die components is generally governed by many factors such as design of die components, selection of materials, heat treatment, techniques of manufacturing of die components, work material, production conditions and the maintenance of press tool components. Failure investigations on numerous worn-out dies from many different applications specify that five main failure mechanisms are encountered of deep drawing tooling. These are wear, chipping, plastic deformation, cracking/total failure, and galling. Figure 1 shows various failure mechanisms in sheet metal tooling.

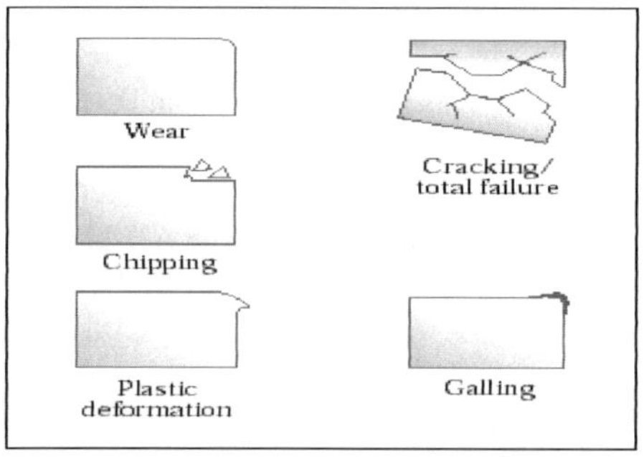

Figure 1. Most Frequent Failure Mechanisms in Sheet Metal Tooling [2]

The wear observed in deep drawing tooling has been classified as abrasive wear, adhesive wear, mixed wear. Abrasive wear dominates when the sheet material is hard and/or contains hard particles such as oxides or carbides. Microscopic examination shows that the tool steel properties that are important for good resistance to abrasive wear are high hardness, high volume of carbides, high hardness of the carbides and large carbide size. Adhesive wear occurs with soft and adhesive sheet materials such as aluminium, copper, stainless steel and low carbon steels. The mechanical properties of die components that are critical for good resistance to adhesive wear are high hardness, low coefficient of friction between the tool and work material, and high ductility. It should be noted that not all sheet materials causes purely abrasive or purely adhesive wear. Some cause partly adhesive and partly abrasive wear. This type of wear is designated as mixed wear. The present research work is planned to investigate the wear behaviour of the deep drawing die tool to characterize the mechanical & tribological properties of die steel and to develop statistical model to predict wear under dry conditions. In this research punch and die work material consider is high carbon, high chromium Viking steel. It offers very high toughness and wear resistance used for manufacturing of punches and dies of deep drawing tooling.

2. Literature review

In this literature review, the various issues pertaining to wear behaviour of tool steels and various testing methods are discussed. Cockeram et al. [3] developed a four promising coating approaches using scratch adhesion and indentation testing for cobalt based alloys. Bahrami et al.[4] investigated the effects of conventional heat treatment on wear resistance of AISI H13 tool steel and effects of in situ surface tempering on AISI H13 tool steel. Lima et al. [5] have made an analysis of AISI D2 tool steel of cold work and AISI 4340 low alloy steel of high strength both of them under hard turning process. Bourithisn et. al [6] have made an investigation using the same hardness (60HRC) of the AISI O1 and AISI D2 tool steels and compared wear properties between them. Rendón and Olsson [7] performed investigation on some commercial abrasion resistant steels to find abrasion wear resistance. Kim et al.[8] established guidelines to select the optimum combination of die materials, coatings and lubricants in stamping galvanized AHSS (DP600, TRIP780 and DP980) for automotive structural parts.

Cora et al.[9] investigated the wear performance of different coatings (two different PVD, CVD and TD coating) applied onto variety of substrate materials (DC 53, SKD 11, DRM 3, DRM 51) against AHSS sheet blanks. A non-reciprocating, CNC-based, slider type of tester was employed in wear tests. Further, Cora et al.[10] developed a new slider type of test system replicate the actual stamping conditions including the contact pressure state, sliding velocity level and continuous and fresh contact pairs (blank-die surfaces). Vilaseca et al.[11] analysed tribological behaviour of two sets of industrial hot stamping tools, one tool made of uncoated hot work tool steel and another PVD coated hot work tool steel set, by means of a surface replication technique. Witulski et al.[12] developed a methodology for the rapid production of drawing tools with high wear resistance to form free-form shaped sheet metal parts. Shih[13] developed a rephosphated steels applying the phosphate coating on zinc coated sheet steels to increase the lubricity in the automotive stamping process and adding extra corrosion protection. He investigated the tribological performance and die wear behaviour of pre-phosphate AHSS in the die tryout and production conditions.

From the reviewed literature it is found that various researches has been done for the wear testing of high carbon high chromium material especially using the AISI D2 grades and experimentation with various coatings. Also these grades of tool steels are not only being performed a comparison wear behavior test amongst each other but also amongst other grades of higher and rougher applications especially used for hot work, plastic moulding and power metallurgy types of applications. The issue still stands since there are extremely less research happening when it comes to high carbon high chromium steels of these grades such as Assab88(Sleipner), Caldie and Viking. Further it is found that most of researchers performed the experimental investigation to determine the wear behaviour of AISI D2 grade steel and proposed few mathematical models based on statistical regression techniques. But research for viking grade material is rarely been done for wear testing purposes. Hence there is a need to study the wear behavior of Viking grade steel. Hence in present study an experimental investigation has been carried out under dry sliding condition to determine wear behaviour of Viking grade steel under different process parameters. The main aim was to study the influences of various process parameters such as applied load, sliding distance, sliding speed under dry sliding wear of the Viking grade steel with the help of factorial experimental design under various testing conditions. Furthermore, the analysis of variance (ANOVA) is employed to examine to test various characteristics of Viking grades steel.

3. Experimental details

The experimentation has been carried out on *Viking grade HCHC tool steel,* it is also known as high carbon high chromium steel. This steel offers very high toughness and wears resistance and used for manufacturing of punches for sheet in copper, stainless steel, zinc, brass, hard abrasive. Further, it also used for other applications such as deep drawing dies, cold extrusion dies, trimmer dies, cupping dies, plug gauges, blanking special taps, broaches etc. The specimen preparation and experimentation were carried out according to ASTM standards.The mechanical properties of Viking grade stainless steel material has a hardness of 58 HRC, yield strength of 1960 N/mm^2, density of 7.75 k/m^3 and modulus of elasticity 190000 N/mm^2. Table 1 gives the chemical composition of the materials used in the investigation.

Table1. Chemical Properties of Viking grade HCHC tool steel

Components	C	Si	Mn	Cr	Mo	V
Wt%	0.5	1	0.5	8	1.5	0.5

Experimental procedure followed for determination of wear is shown in Figure 2. Before start of all tests, the specimens and holders were thoroughly cleaned in petroleum spirit in an ultrasonic cleaner, thereafter rinsed with ethanol and dried in air.

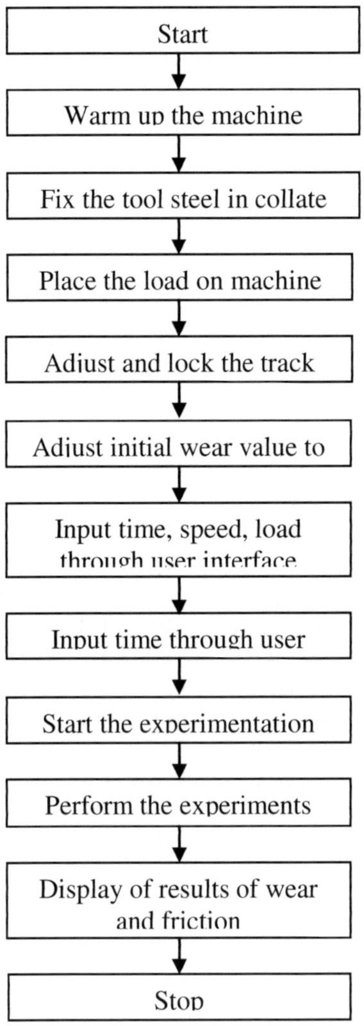

Figure 2. Experimentation Flow Chart.

For each of the speed of the rotating disc these three loads will be experimented on the pin. The track diameter is considered to 60 mm and 75mm and duration of each of the experiment is considered to be 10 minutes. The output parameters to be obtained from the experimentation are the wear rate of the tool steel in micro-meters/meter and the coefficient of friction of the tool steel. Also the minimum wear rate of the tool steel, the maximum wear rate of the tool steel and the mean or the average wear rate of the steel is also considered. Initially the machine has to be switched on and the warming up of the rotating disc has to be performed at around 500 rpm for 1 to 2 minutes. After the automatic stopping of the warm-up rotation of the disc, the track diameter which was considered is now set up, screwed and locked.

At the end of the arm is where the pin or the tool steel is being held and it consists of collate. There are various shapes and sizes of collates depending upon the shape and size of the tool steel. A squared shaped with a dimension of 10 mm x 10 mm tool steel is being used for experimentation. The tool steel is placed in such a way that there is a sufficient hard contact between the tip of the pin and the rotating disc without any load and the pin is then fixed and screwed tightly. Lastly for desired input parameter, the desired load is then placed. Now the duration of the time is set which is 10 minutes and the initial wear value of the pin is set to a tolerance limit between +30 to -30 by fine tuning of the two sensitive lead screws. The DUCOM Micro POD pin-on-disc tribometer is interface with automatic data acquisition system. Initially, through user interface all input parameters such as speed, load, and duration and track diameter enter into system. During experimentation the system recorded wear behaviour and coefficient of friction readings of the tool steel on computer screen as shown in Figure 3.

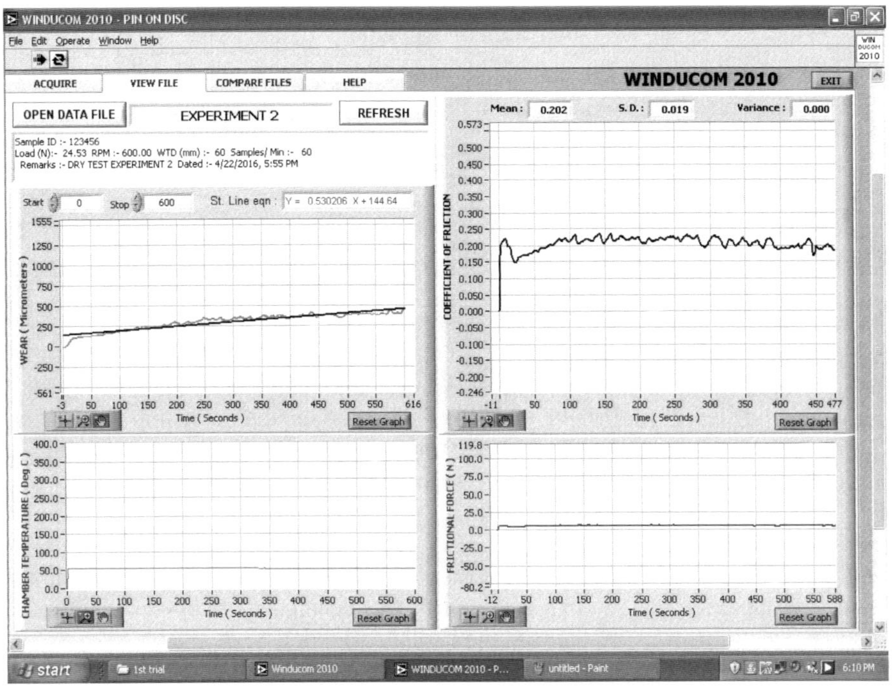

Figure 3. Screen shot of acquisition of wear behaviour and coefficient of friction.

4. Test equipment

For the wear testing of Viking grade using Pin-On-Disc tribometer, the DUCOM Micro POD Pin on Disc machine is being utilized is shown in Figure 4.The Ducom Micro PoD comprises of a precision controlled drive that controls disc rotation. The counter specimen, which may be a ball, pin or any other custom geometry, is held firmly in a friction/loading arm. A known load is applied upon the counter specimen to achieve the desired contact stress. The Micro PoD is a fully computer controlled

IOP Conf. Series: Materials Science and Engineering **244** (2017) 012004 doi:10.1088/1757-899X/244/1/012004

test system. The operator may choose and set the desired test parameters through the user interface provided on DucomTribo.

Figure 4. DUCOMMicroPOD pin-on-disc Tribometer.

5. Design of experiments

The most important stage in the plan of experiments is selection of factors.The input parameters that are considered for this experiment are the speed of the rotating disc and the load which will be transferred to the stationary pin that applied on the rotating disc. The levels of the input parameters for the rotating disc are considered to be 600 rpm, 900 rpm and 1200 rpm and for the load it is considered to be 1.5kg, 2.5kg and 4.0kg. For each of the speed of the rotating disc these three loads will be experimented on the pin. The track diameter is considered to 60 mm and 75mm and the duration of each of the experiment is considered to be 10 minutes. Table 3 indicates the process parameters and their levels.

Table 2. Process parameters and their levels

Factors	Levels		
	I	II	III
Load (N)	1.5	2.5	4.0
Speed (rpm)	600	900	1200
Track Dia. (mm)	60	--	75

Three level factorial experiments create to investigate the effect of load, speed and track diameter on wear of Viking steel. About 27 experimental runs were conducted by including the effect of randomization and replication. The experimental results were analyzed using analysis of means and variance tostudy the influence of factors. The Experiments were conducted as per the standard factorial design so as to investigate which design parameter significantly affects the dry sliding wear for the selected combinations of load, sliding speed and sliding distance and material.

6. Results and discussions

In orderto identify the main controlling parameter(s) which having influence on the wear behaviour of Viking grade steel during operation, an experimentation has been carried out on DUCOM Micro POD pin-on-disc Tribometer. The observations from the investigation have been summarized and discussed in following section.

6.1. Influence of load and speed on the wear behavior

The load has a significant influence on the wearloss of the material. Figure 5 represents the variations of wear rate of the Viking gradesteel as a function of applied load for a constant sliding distance 65mm and speed 700rpm, 900 rpm, 1200 rpm at normal room temp. It may be noted that wear rate of Viking gradesteel metal increased with increasing load. The wear rate of metal is depended on the heat treated conditions of the metal and the applied load. During this process, transfer of pin material to the disc was also observed. This type of seizure has been referred to as galling seizure which leads to further increase in wear rate. Wear rate of normal specimen increases gradually with applied load. The wear rate of the pin increases with increasing speed. As the speed increases the temperature increases which leads to plastic deformation ofthe material. At lower speeds the pin surface experiences severe damage resulting in a high wear rate. It can be noticed the wear rate increases with the increase in speed. If the normal specimen is considered the wear is increasing continuously upto 900 rpm beyond which it becomes stable up to 1000 rpm due to the formation of the smeared layer and wear rate start reducing continuously from 1050 rpm to 1200 rpm. Figure 5 shows the surface plot for wear rate and effect of speed and load on wear of Viking grade steel and the interaction plot between speed and load on wear is shown in Figure 6.

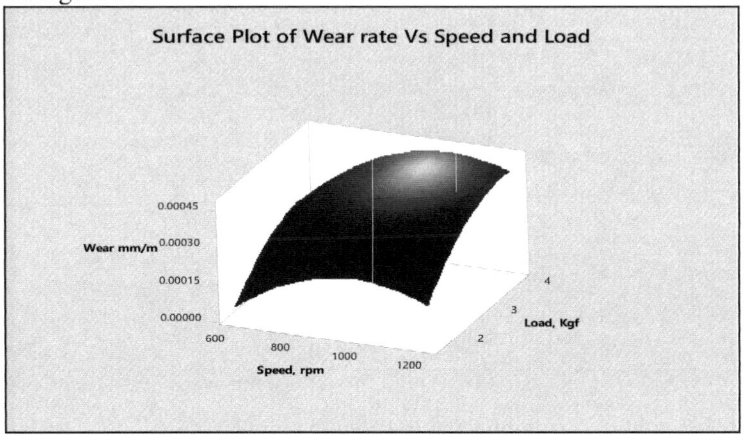

Figure 5. Surface plot for wear rate and effect of speed and load.

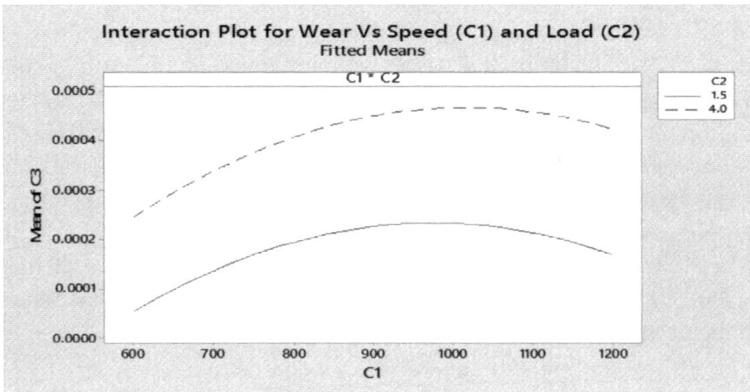

Figure 6. Interaction plot between speed and load on wear.

7. Analysis ofvariance (ANOVA)

The ANOVA is used to investigate which design parameters significantly affect the quality characteristics. It is accomplished by separating total variability of S/N ratio, which is measured by sum of square deviation from total mean S/N ratio into contributions by each of the design parameters and the errors.

Table 3. Regression Statistics.

Regression Statistics	
Multiple R	0.9175
R Square	0.9437
Adjusted R Square	0.9372
Standard Error	1.10185

The above table 3 shows that the R square value obtained is 0.9437 and adjusted R square value is 0.9372 and this enlighten us that the model account for only 94.37% of variance in the output. This is the clear indication that this model is a statistically good model.

Table 4. ANOVA table

Source	DF	Adj SS	Adj MS	F-Value	P-Value
Regression	5	477.197	95.4394	78.61	0.000
Load (C1)	1	31.917	31.9174	26.29	0.000
Speed (C2)	1	4.382	4.3825	3.61	0.071
C1*C1	1	90.288	90.2876	74.37	0.000
C2*C2	1	0.345	0.3452	0.28	0.599
C1*C2	1	68.101	68.1007	56.09	0.000
Error	21	25.496	1.2141		
Lack-of-Fit	3	25.496	8.4986	*	*
Pure Error	18	0.00	0.000		
Total	26	502.692			

From the above ANOVA table 4 it is inferred that the percentage of each factor contribution on the total variation and thus exhibiting the significant influence on the wear of the steel. The contribution of applied load is greater which may be due to the sufficiently induced stress at contact area within the experimental conditions. The linear regression equation for the wear loss of the Viking steel is shown in equation below:

Wear (C4) = 12.58 - 0.04775 C1 + 3.53 C2 + 0.000043 C1*C1 + 0.161 C2*C2 - 0.006311 C1*C2

The above equation gives a measure of contribution of each variable to the model. The coefficient indicates that for addition unit of C1 the wear rate is expected to increase by 0.04775 unit. Thus, load (C1) is influencing the wear rate. It indicates that load is the main factor on wear rate for Viking steel. It is followed by sliding speed while the sliding distance was less effective than the other parameters. The coefficients of applied load, sliding distance, sliding speed and material conditions are positive, which indicates that wear increases with the increase in the wear parameters. It indicates that load is the main factor on wear rate for Viking steel. It is followed by material conditions and sliding speed while the sliding distance was less effective than the other.

8. Conclusion

Based on experimental analysis carried out on the Viking grade tool steel and obtained results, it is concluded that the processing parameters that are considered for the wear operations are significant and also these parameters are significant amongst each other. Wear rate of Viking grade tool steel material increased with increasing load and speed. Since the analysis of this tool steel is of non-heat treated tool steel under dry test at normal temperature. Hence future scope would be analysis of wear behaviour of a heat treated material under dry run as well as under various lubricating oils at different working temperature. Further analysis of wear resistance and the wear behaviour of the coated and uncoated Viking grade under various lubricating oils are also some of the future scope to be considered.

References

[1] Naranje V and Kumar S (2014) Expert Systems with Applications, 41, 4, 1419
[2] Kumar S (2011) Journal of Engineering Research and Studies, 2,119
[3] Cocheram B V, Wilson W L, (2001) Surface and Coatings Technology 139, 161

[4] Bahrami, A, MousaviAnijdan SH, Golozar MA, Shamanian M & Varahram N (2005) Wear,258, 5–6, 701
[5] Lima J G, De Ávila R F, Abrao A M, Paulo Davim J (2005), Journal of Materials Processing Technology, 169(3), 388
[6] Bourithis L, Papadimitriou G, Sideris J (2006), Tribology International 39(6):479-489 · June 2006
[7] Rendón J and Olsson M (2009), Wear 267(11):2055
[8] Kim H, Han S, Yan Q, Altan T (2008)CIRP Annals - Manufacturing Technology, 57 299
[9] Cora O. N, Ali A, Namiki K, Sofuoğlua H, Koçb M (2012),Tribology International 52, 50
[10] Cora O. N, Koç M, (2009), Wear 267 (2009) 1123–1129
[11] Vilaseca M, Pujante J, Ramírez G, Casellas D (2013),Wear, 308,1-2,148
[12] Witulski J, Trompeter M, Tekkaya A, Kleiner M, (2014) CIRP Annals- Manufacturing Technology, 60 (1),311
[13] Shih (2016)- Shih, H., (2016), SAE Int. J. Mater. Manf. 9(2):481

Influence of cryogenic cooling rate on mechanical properties of tool steels

G Mazor, I Ladizhensky, A Shapiro

Research Center for Energy and Numerical Simulations (RCENS), Department of Mechanical Engineering, SCE - Shamoon College of Engineering, Beer-Sheva 84100, Israel
Email: mazorg@sce.ac.il

Abstract. The effect of the rapid cryogenic treatment on hardness and wear resistance of several kinds of tool steel was examined. Two ways of cryogenic cooling were evaluated: direct immersion of the metallic samples into liquid nitrogen and three-stage rapid cryogenic cooling (1 – precooling in LN_2 to -20°C, 2 – formation on the sample of a frost layer from air by natural humidity, 3 – second cooling of the frost-covered sample in LN_2 to -195.7°C). Material in "as is" conditions and after a preliminary heat treatment (850°C) were used as the reference points.
The HV microhardness and the wear rate under dry abrasive friction were evaluated. Despite the very different types of the examined metals' nature, microstructure, and hardening mechanisms, the rapid cryogenic cooling improves both the hardness and the wear resistance values. For all investigated metals rapid cryogenic cooling assisted with the frost layer produces the best results.

1. Introduction

Improving mechanical properties of metallic materials, particularly tool steels, is an important problem for metallurgy, metals machining technology, metal treatment, mechanical engineering, and other fields. One of the technologies used to increase the workability of tool steels and other metallic materials is cryogenic treatment.
It is known that cooling rate is a very important characteristic of the quenching of bodies in contact with boiling cryogenic fluid. The adjustment of the cooling rate for individual process requirements allows achieving the needed material characteristics [1, 2]. For example, shortening the cryogenic treatment period for steel samples causes considerable improvement of the steel toughness, bending strength, etc. In order to reduce the time to achieve the desirable temperature, it is necessary to intensify the heat transfer between the items that undergo cryogenic treatment and the boiling cryogenic fluid.
Lastly, we have studied the following types of short-term boiling amplifiers of cryogenic liquids: frost layers [3, 7], salt produced coatings, and the coatings formed by dried foams or detergents [4]. The mentioned porous layer on the surface of an intensively cooled body suppresses formation of large steam bubbles, decreasing heat exchange between the cooling liquid and the body.

2. Materials and methods

2.1. Experimental setup

Experimental setup for a combined heating-cooling treatment includes (Figure 1): two-end open vertical tube furnace ELF 11/14B (by Carbolite) with temperature controller; the heated volume is ∅40×300 mm and a maximal temperature of 1000°C; mechanisms for rotating, lifting, and lowering the sample, water or salt solution sprayer (atomizer); Dewar vessel with liquid nitrogen (∅170×254 mm internal size); T-type thermocouple and data acquisition system.

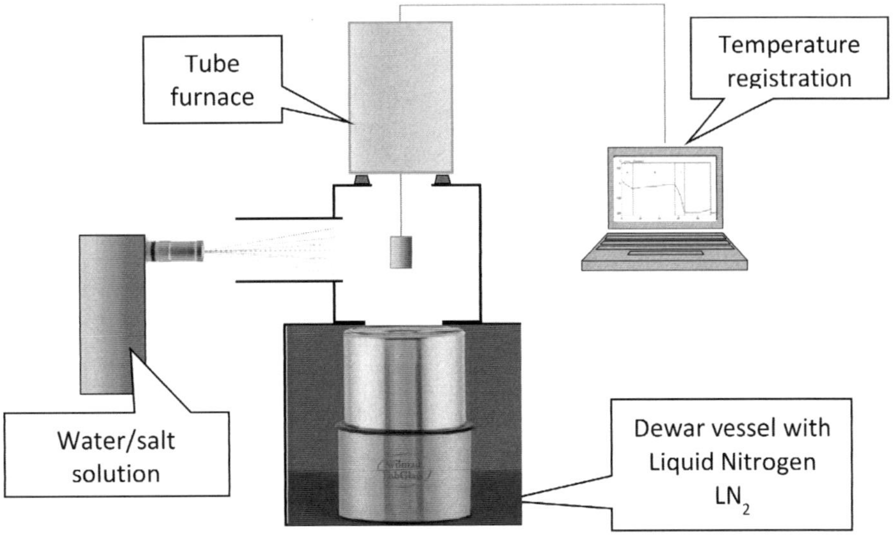

Figure 1. Experimental setup for combined heat treatment.

An aqueous solution of magnesium sulfate ($MgSO_4 \cdot 7H2O$) of 50% concentration was used in the experiments. The sample was fixed in a sample rotation unit and heated to 150-200°C. The salt solution, heated to 100°C, was sprayed on the sample surface during one rotation period. The preheated metallic sample evaporates the water quickly. The salt solution sprayer enables creating a rather uniform salt layer on all surfaces of the treated sample.

2.2. Materials and samples

In our work we investigated different kinds of steels:

- AISI D2 (high-carbon, high-chromium cold work steel;
- AISI O1 (oil-hardening cold work steel);
- AISI H13 (chromium hot work steel).

The steels are of very dissimilar types, and industrially they are exposed to rather different heat treatment procedures including annealing, preheating, and austenitizing temperatures, quenching media, etc.

The steel samples were subjected to two different testing procedures: microhardness and abrasive wear rate evaluations. All measurements were made on 15 mm length and ∅15 mm diameter cylindrical specimens. The Vickers microhardness and the wear rate were measured on the opposite flat basic surfaces of the sample. For the specimen temperature measurement, the T-type thermocouple was placed into the geometric center of cylinder.

2.3. Treatment sequence

Our main goal was an evaluation of the effect of cryogenic cooling rate on different metallic materials' wear rate. Two series of experiments were performed.

IOP Conf. Series: Materials Science and Engineering **244** (2017) 012005 doi:10.1088/1757-899X/244/1/012005

The first series was dedicated to evaluation of dependency of tool type high carbon high chromium steel AISI D2 wear resistance on different kinds of temperature treatments including cryogenic cooling (CC) and rapid CC (RCC).

A number of heat treatment and cooling schemes were tested. The maximal variation of cooling rate values did not exceed limit of 17%.

Schematic steel temperate evolution curves are demonstrated on Figure 2.

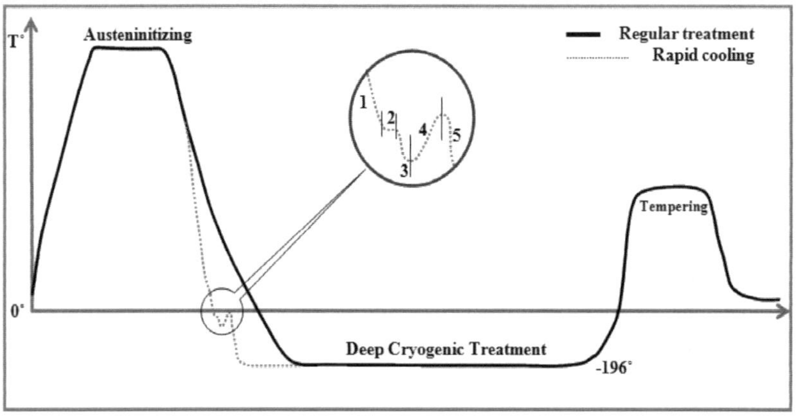

Figure 2. Scheme – regular heat treatment (includes CC) vs. RCC.
1 – water or salt solution spraying, 2 – sample drying, 3 – cooling to -20°C, 4 – frost layer forming, 5 – final rapid cryogenic cooling.

Table 1. Heat treatments and cooling rates.

Regime	Materials	Heating temperature, °C	Heating Duration, min	Sample coating	Maximal cooling rate, °C/s
1	D2, H13, O1	No heating	No heating	No coating	2.1
2	D2	700	40	No coating	4.9
3	D2	850	55	Salt - $MgSO_4 \cdot 7H_2O$	5.7
4	D2	850	55	Frost (by water spray)	6.0
5	D2, H13, O1	850	55	No coating	4.7
6	D2, H13, O1	850	55	Frost (by moist air)	8.4

Figure 3 shows surface of no treated sample (Figure 3a), thick and preventing effective heat exchange frost layer (Figure 3c) and optimal sample state. (Figure 3b).

Optimal thickness of the frost layer (Figure 3b) was determined in a number of preliminary experiments under different moisture and temperature conditions.

Duration of each cooling stage for regimes 3-5 was established in a number of preliminary experiments.

At second stage (with steels H13, D2 and O1) of investigation the number of heat treatment regimens was reduced down to numbers 1, 5 and 6 Table 1).

Figure 3. The frost layer evolution: a – initial sample surface; b – frost layer (black region) created from moist air on cooled sample surface. First, fabricated needle-like structures increase the contact surface value and help to intensify heat exchange; c – thick and dense coverage isolates the sample from contact with cooling media.

2.4. Sample testing methods

The temperature of the sample was measured by a T-type thermocouple attached in the center of the cylinder and connected to the data acquisition system with PycoLog software for data processing and saving. Prior to any measurement, the thermocouple was calibrated by comparison with a Standard Platinum Resistance Thermometer.

The samples' microhardness was measured by microhardness tester HVS-1000B (MITECH) under 10 N load. Wear rate and microhardness measurements were performed on different sample surfaces. Microhardness of samples was measured in accordance with ASTM E92 Standard. Wear rate evaluation was performed by a homemade pin-on-disk abrasive wear tester in accordance with ASTM E290 Standard and a specific wear rate value W_s was calculated.

3. Results and discussion

3.1. D2 steel – different heat treatments

Results of the first experiment on D2 steel are shown on Figure 4. Steel D2 in as-is condition has a relatively low microhardness of 253 MPa, which decreases after heating at 700°C and increases if temperature reaches 850°C. However, both levels of heating increase the wear rate of steel. Regular cryogenic cooling of steel has no effect on steel properties without preliminary heating and even may slightly enlarge the wear rate value (Figure 4, As Is + LN2). Any combination of steel preheating followed by cryogenic cooling decreases the D2 steel wear rate. The most significant effect demonstrates heating of steel to 850°C followed by frost layer-assisted cryogenic cooling. In this case we observe decreasing of the wear rate to 50% of the initial level accompanied by doubling of the microhardness value of the steel.

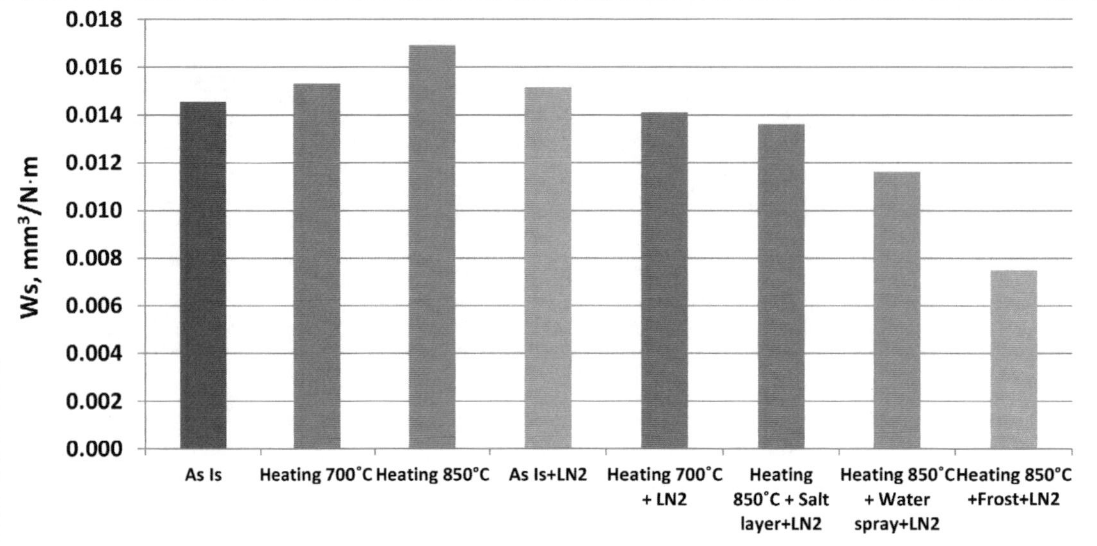

Figure 4. Wear rate of D2 steel under different heat treatment conditions.

3.2. Tool steels – Cryogenic cooling vs. rapid cryogenic cooling

All experimental data presented in Figures 4 and 5 prove that the most effective method of decreasing metals' wear rate is a combination of suitable heating and rapid cryogenic cooling of the metal. The method works for all kinds of steels tested.

Results of the second experimental stage on different steel can be seen on Figure 5.

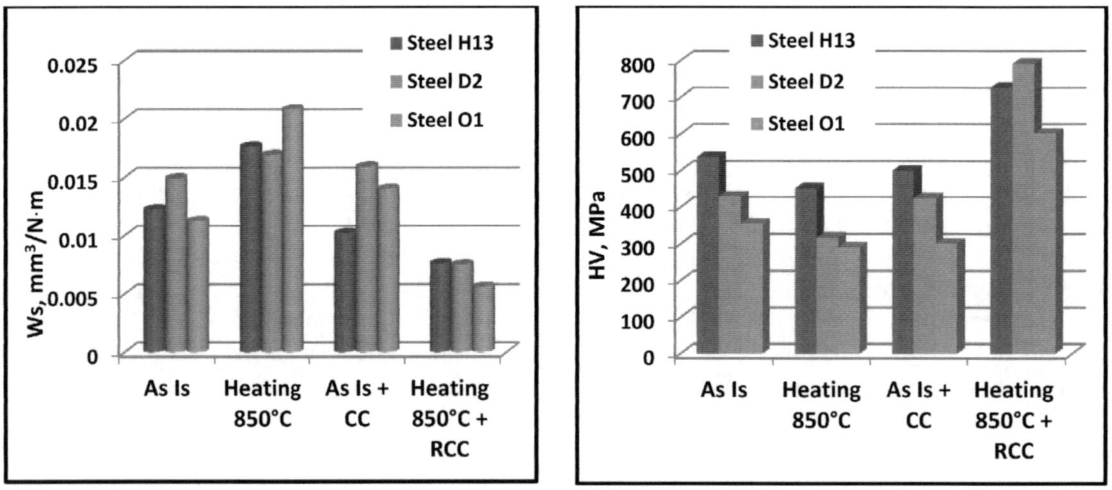

Figure 5. Wear rate and microhardness of tool steels. The best result was achieved by a combination of preliminary heating to 850°C and rapid cryogenic cooling.

It's important to note that the critical influence of non-stop cooling treatment steels starting from the high tempering temperature down to the temperature of liquid nitrogen were stated in the earliest studies of cryogenic treatment of steels [5, 6]. Authors, using different experimental techniques, showed that any delay in placement of steel samples into liquid nitrogen causes an increase of retained austenite quantity. Sometimes this austenite cannot be transformed to martensite by longtime cryogenic cooling.

There are several ways to explain improving of cryogenically treated steels' resistance to abrasive wear:1) first of all, this reduces the retained austenite quantity. Continuous cooling from 850°C to -195.7°C helps to increase the martensite quantity and properties and to produce a finer internal microstructure of treated steels. It is known that intensive plastic deformation of steel leads to the austenite to martensite transformation. So, the bigger temperature gradient developed by rapid cryogenic cooling and accompanied by intensive plastic deformation may decrease retained austenite present and improve such mechanical properties of steel as toughness and wear resistance; 2) another mechanism increasing steel's mechanical properties is creation of finely dispersed carbides precipitation. As was noted in [2] that fine carbide precipitation appears only in case of deep cryogenic cooling under -70°C.

It was not revealed any samples' micro-cracking created by swift cryogenic cooling of steel from very high temperatures. It may be partially explained by different specific volumes of austenite and martensite: martensite has a larger elementary atomic cell than austenite and this expansion of steel partially compensates for internal stress resulting from the significant temperature gradient in the surface layers. The large internal strain established by rapid cooling helps with additional austenite to martensite transformation and to creation of a highly dispersed internal structure that reduces residual stresses and prevents the treated steel from cracking. However, any recommendation for cryogenic treatment of specific steel or detail requires scrupulous examination of the material state at each treatment step.

Actual papers dedicated to deep cryogenic treatment of steels recommend using of rather low cooling rates of a few Kelvins per minute [8]. So, cryogenic treatment takes a long time. However, in accordance with our results, some small dimensions details like cutting tools may be treated at the higher cooling rate without microcracking damage.

Detailed investigation of the observed influence of cryogenic cooling rate on steel mechanical properties may be achieved in the future by microscopic analysis of the treated steel structure.

4. Conclusions

1. As discussed in the above experimental results, the frost layer assisted rapid cryogenic cooling of the investigated metal together with preliminary metal heating, creating the lowest wear rate of treated metals. This conclusion is correct for the studied ranges of cooling rates.
2. Discussed rapid cryogenic cooling method is rather universal and may be used for treating of very different types of metallic materials.
3. Further research in improving of metals' wear resistance must be concentrated in following directions:
   - Optimization of metal preliminary heating conditions, such as temperature and duration of the treatment.
   - Evaluation of post-cooling treatments' (tempering) effect on workability of metal.
   - Optimization of frost layer forming conditions to create a uniform frost layer with increased heat exchange possibility.

5. References

[1] Kamody DJ Cryogenics process update 1999 *Adv Mater Process* **155**(6):H67–9
[2] ASM Handbook 1991 Heat treating. Materials Park, OH: ASM International
[3] Mazor G, Korin E, Nemirovsky D, Ladizhensky I 2013 Frost formation as a temporary enhancer for quench pool boiling *Appl Therm Eng* **52**(2):345-52
[4] Ladizhensky I, Korin E, Mazor G, Nemirovsky D, Goldkin E 2014 Quench pool boiling with temporary crystalline enhancers. *Chem Eng Technol.* **7**(2):349 56
[5] Collins DN Deep cryogenic treatment of tool steels, A Review 1996 *Heat Treat Met*. **2**:40-2
[6] Baldissera P, Delprete C 2008 Deep Cryogenic Treatment. *Mech Eng* J. **2**:1-11

[7] Ladizhensky I, Mazor G, Korin E, Nemirovsky D, Alfi A, Rabin A and Yehud E 2017 Accelerated cryogenic cooling caused by the frost layer enhancer. *Journal of Heat Transfer* ASME, **139** (2) , pages 7 http://heattransfer.asmedigitalcollection.asme.org/article.aspx?articleID=2566023#signInButton Section
[8] Ray KK & Das D Improved wear resistance of steels by cryotreatment: the current state of understanding, *Materials Science and Technology*, 2017, **33**(3) pp.340-354

IOP Conf. Series: Materials Science and Engineering **244** (2017) 012006 doi:10.1088/1757-899X/244/1/012006

Comparison of tool life and surface roughness with MQL, flood cooling, and dry cutting conditions with P20 and D2 steel

S W M A I Senevirathne[1] and H K G Punchihewa[2]

1, 2 Department of Mechanical Engineering,
University of Moratuwa
Moratuwa, Sri Lanka
amal@mech.mrt.ac.lk , hkgpunchihewa@gmail.com

Abstract. Minimum quantity lubrication (MQL) is a cutting fluid (CF) application method that has given promising results in improving machining performances. It has shown that, the performance of cutting systems, depends on the work and tool materials used. AISI P20, and D2 are popular in tool making industry. However, the applicability of MQL in machining these two steels has not been studied previously. This experimental study is focused on evaluating performances of MQL compared to dry cutting, and conventional flood cooling method. Trials were carried out with P20, and D2 steels, using coated carbides as tool material, emulsion cutting oil as the CF. Tool nose wear, and arithmetic average surface roughness (Ra) were taken as response variables. Results were statistically analysed for differences in response variables. Although many past literature has suggested that MQL causes improvements in tool wear, and surface finish, this study has found contradicting results. MQL has caused nearly 200% increase in tool nose wear, and nearly 11-13% increase in surface roughness compared flood cooling method with both P20 and D2. Therefore, this study concludes that MQL affects adversely in machining P20, and D2 steels.

1. Introduction

Cutting Fluid (CF) or Metal Working Fluid (MWF) are highly important in metal cutting, as these fluids provide three main functions in machining: Primary function- heat transfer from the cutting area; Two secondary functions- providing lubrication in cutting area, and removing the debris from the cutting area. Most common type of fluid used in metal cutting is emulsion CF [1]. However, the expenditure for cutting fluid in the European automotive industry is to be nearly 20% of the total manufacturing cost [2]. According to the same source, the cost of tools was only 7.5% of the total manufacturing cost, making the cutting fluid cost to be comparatively high than other costs associated with manufacturing. Typically, metal cutting operations are generating a great amount of heat, therefore it is obvious that most of the energy consumed by the operation is converted into heat [3]. If not controlled properly, this heat will affect the cutting system adversely, leading in to poor performances in cutting systems, and manufacturing defects as well [3]. Thus, effective heat dissipation from the working area is essential to bring down manufacturing defects, and improve performances in the cutting systems such as reducing tool wear, reducing surface roughness of machined parts, etc. There are many CF application methods developed in industry such as flood cooling and high pressure cooling [1]. Development on material has demanded more performance from these CFs and CF application methods, but found to be insufficient for machining modern material. There are newer methods developed, for example, cryogenic machining and minimum quantity lubrication (MQL), in the recent history [1]. However, under numerous circumstances, these newer methods have also shown limitations [4], [5]. Previous studies have shown that, both cryogenic machining and MQL machining have shown promising results [6][7]. However, cryogenic machining has several disadvantages over MQL machining such as higher cost and potential safety hazards.

In MQL a minute amount of CF is mixed with high-pressure air and the resulting aerosol is directed to the cutting area through a nozzle that flutters aerosol to the cutting area at high speed [8], [9] [10]. The flow rate of CF is typically 50 – 400 ml/hour [4], [11]–[15]. While air in the aerosol provides the cooling function and chip removal, the liquid provides lubrication and by droplet evaporation, it also provides cooling [16]. Primarily, the CF will absorb heat from the cutting area due to evaporation, and in addition, to a minor extent, by vaporisation also [17]. MQL method does have several advantages over the other CF application methods. Since the amount of CF being used is low, the chip, work-piece, and the tool holder will have low amount of fluid residue after the machining operation making cleaning function easier and cheaper [17]. Also the cutting operation can be easily observed as the CF is not obstructing the view as in flood cooling method. However, previous research shows that the cooling capacity of MQL is depending on the work-tool material combination. The effect of MQL on machining AISI P20 and D2 tool steels has not been previously done. However, these two tool steels are popular in the tool making industry, and hence warrants an investigation.

2. Methodology

A custom designed MQL aerosol application system which comprises of refrigeration system and CF aerosol distribution system, was used for the experiments. After reviewing previous literature, an initial set of parameters were selected for the experiment, and further refined after a pilot run. The response variable was the wear of the tool tip, and the arithmetic average surface roughness (Ra) of the machined work-piece. The results of the pilot study were also used to further refine the experiment parameters. A two level factorial experiment was designed [18]. Work-piece material was taken as the level one with two types of material, i.e. AISI P20 and AISI D2. Three cooling conditions, i.e. dry cutting, conventional flood cooling, and MQL, were taken as the second level.

During the trials, parameters were controlled as given in Table 1. In order to enable comparison of results with past research, and ensure simplicity of the machining operation and ease of operation, straight turning operation was selected for the study. Feed rate and depth of cut were selected based on the tool manufacturers' recommendation and by reviewing literature. Spindle speed was also chosen following the tool manufacturers' recommendation based on the surface cutting speed. Length of the cut was determined by the results of the pilot run. MQL aerosol flow rate, pneumatic pressure, nozzle target location and nozzle direction was selected through a literature review.

Table 1. Controlled parameters for the experiments.

Parameter	Value	Parameter	Value
Turning Operation	Straight	Flow Rate (MQL)	160 ml/ hour
Feed Rate	0.5 mm / rev	Pneumatic Pressure	7 bar
Depth of Cut	0.5 mm	Flow Rate (Flood)	9 l/min
Length of the Cut	500 mm	Nozzle Target Area	Rake Face
Surface Cutting Speed	150 m/min	Nozzle Angle (H)	$\approx 60^{o}$
Spindle Speed	1220 rpm	Nozzle Distance	150 mm

2.1. Experimental setup

A 3.0 kW horizontal manual lathe machine was used for the straight turning operation. A total of 18 work-pieces with 9 pieces from AISI P20 and 9 pieces from AISI D2 tool steels were used for the trials. The cylindrical work-pieces were of 45 mm diameter with 100 mm length. The cutting length of 500 mm was obtained by 10 passes of 50 mm. Chemical Vapour Deposition (CVD) coated carbide turning tool from a popular commercial brand was used to machine both types of work-piece material.

The coating on the tool was TiCN-Al_2O_3-Ti with a hardness of 90.3 HRA. The prepared CF had a ratio of 1:9 emulsion oil to water by volume, followed by the CF manufacturers' recommendation. The orientation of the aerosol was in a vertical plane perpendicular to the axis of rotation of the work piece with an upward inclination to the horizontal by 60° and the aerosol was directed to the rake face of the tool as shown in Figure 1. The MQL aerosol nozzle was calibrated to spray 160 ml of CF per hour with a pneumatic pressure of 7 bar. The orifice diameter of the nozzle was 1.2 mm. calibration of the nozzle was performed by measuring the time taken to spray a measured quantity of CF. After each 500 mm long simple turning operation, both the workpiece and the cutting tool were coded to enable identification. Afterwards the workpiece was wrapped in an airtight plastic container and immediately taken to measure the roughness of its machined surface. This process was repeated for all 18 workpieces to minimise the effect of corrosion and contamination of the machined surface.

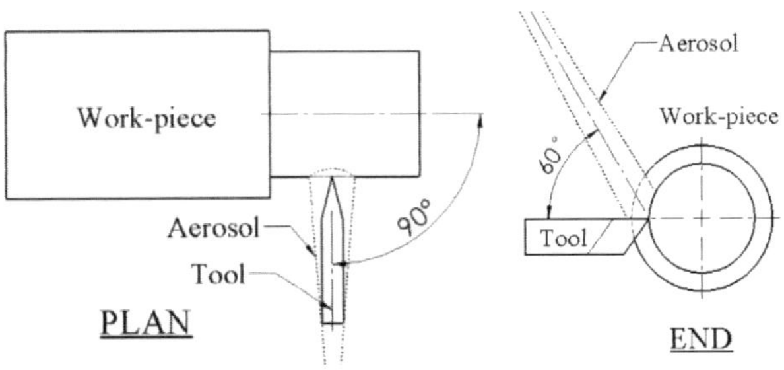

Figure 1. Target direction and orientation of the aerosol.

2.2. Response variables

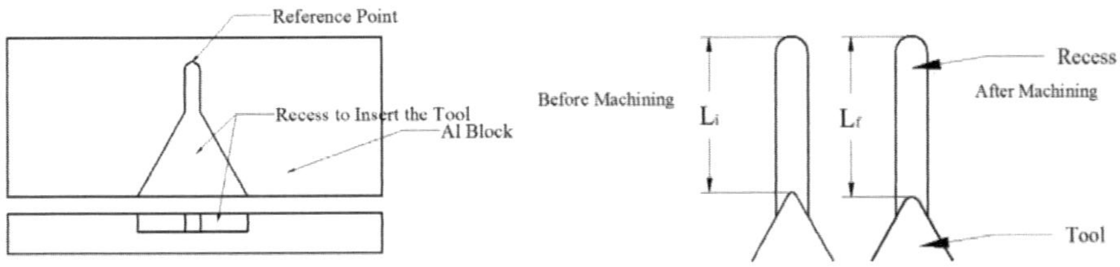

Figure 2. Tool nose wear was measured by the difference in the measured distance from the tool tip to a reference point in a measuring jig.

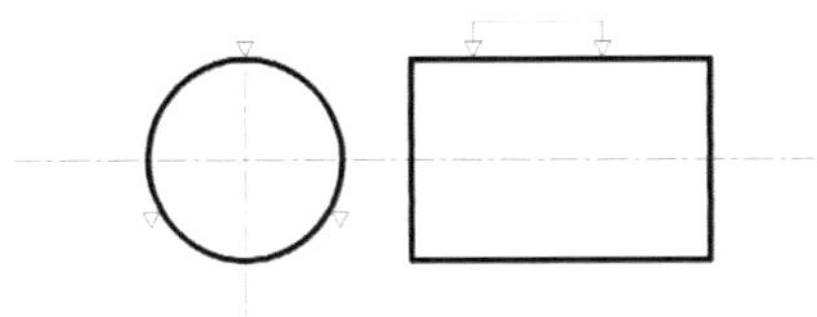

Figure 3. Surface roughness measurement locations and direction.

A CNC-CMM, was used for measuring tool tip radius, and distance between tip and reference point in the specially made measuring jig as shown in the Figure 1. The CMM has a resolution of 0.0001 mm. Since tool nose radius readings were given unexpected results, it was decided to use direct nose wear as the indicative measure for tool wear. A measuring jig was prepared by measuring several samples of unused tools and creating a recess in a steel block as shown in Figure 2, using a CNC machining centre. Used tool was inserted in this block and distance between the worn nose and reference point in the jig was measured as illustrated in Figure 2. Tool nose wear was found by taking the difference in tool nose and reference point distance with used tool and unused tool. A digital surface roughness tester was used to obtain the surface roughness of the machined surface. It directly read the arithmetic average surface roughness (Ra) along the measured distance to an accuracy of 0.01 µm and a cut-off length of 2.5 mm. The surface roughness was measured parallel to the principal axis, across the lay at three different locations on the periphery of machined cylindrical surface, which were 120° apart, as shown in Figure 3. All measurements were repeated thrice.

3. Analysis

3.1. Statistical significance of differences in groups of tool nose wear and Ra
One-way ANOVA was done for each P20 and D2 groups separately for each response variable, tool nose wear and Ra. Tool nose wear readings and Ra readings of three groups of cooling condition as shown in Table 2 with an equal sample size and repeated for each work material. Samples from dry cutting, flood cooling, and MQL were taken for each work material as the groups. A null hypothesis "Mean tool nose wear / Ra of all groups are equal" with an alternative hypothesis "At least one mean tool nose wear / Ra is different" was used for the test.

Table 2. Groups for ANOVA of tool nose wear and Ra.

	Tool nose wear			Ra		
Group	Dry	Flood	MQL	Dry	Flood	MQL
P20	✓	✓	✓	✓	✓	✓
D2	✓	✓	✓	✓	✓	✓

3.2. Statistical significance of differences in between mean Tool wear groups and in between mean Ra groups
Single-tailed Student's t-test was used to compare means of tool nose wear Ra of work-pieces in each cooling condition using 95% confidence level for each of the work material separately. Mean Ra of each group was tested against the rest to evaluate the difference in means Ra is significant or not, and high or low if the difference is found to be significant. Pairs of mean Ra of groups were taken as shown in Table 3. This was repeated for both work material for mean tool nose wear and mean Ra.

3.3. Effect of work material on Ra
Single-tailed Student's t-test was used to evaluate the significance of mean tool nose wear and mean Ra of each cooling condition in each material against the corresponding group of the other material. The pairs for the Student's t-test is shown in Table 3.

Table 3. Pairs of mean tool nose wear and mean Ra for Student's t-test.

Same material groups

Group / Group	Dry	Flood	MQL
Dry		×	×
Flood			×
MQL			

Between P20 and D2 groups

		P20		
		Dry	Flood	MQL
D2	Dry	×		
D2	Flood		×	
D2	MQL			×

4. Results

4.1. Tool nose wear and Ra in AISI P20 and D2
Mean Tool nose wear calculated with the three different cooling methods, and Ra measured for AISI P20 and D2 is shown in Table 4. With each work material, the mean tool nose wear measured with MQL machined tools was observed to higher than that of dry cutting and flood cooling methods. Further, the mean tool nose wear in tools machined with each work material seen to be similar. Mean of arithmetic average surface roughness measured in each steel is shown in Table 4. P20 work-pieces had its highest mean Ra with MQL while D2 had the highest with dry cutting. Lowest mean Ra with

either work material was seen with flood cooling. Patterns of Ra through dry cutting, flood cooling, and MQL were dissimilar for the two materials

Table 4. Tool nose wear and Ra results for AISI P20 material.

Cooling Condition	Mean Tool Nose Wear (mm)		Mean Ra (μm)	
	P20	D2	P20	D2
Dry	0.0643	0.0557	1.514	1.623
Flood	0.0283	0.0299	1.406	1.373
MQL	0.0867	0.0907	1.594	1.518

4.2. Analysis of Variance (ANOVA)
One-way ANOVA with a confidence level of 95% resulted following conclusions. P20 tool nose wear groups concluded with an F-value of 33.37, and P-value of 0.001 (<0.05). Hence it is concluded that at least one mean tool nose wear of the tested three groups of cooling condition is significantly different from the others. D2 tool nose wear groups concluded with an F-value of 42.12, and P-value of 0.000 (<0.05). Hence it is concluded that at least one mean tool nose wear of the tested three groups of cooling condition is significantly different from the others. P20 Ra groups concluded with an F-value of 308.32, and P-value of 0.000 (<0.05). Hence it is concluded that at least one mean Ra of the tested three groups of cooling condition is significantly different from the others. D2 Ra groups concluded with an F-value of 251.87, and P-value of 0.003 (<0.05). Hence it is concluded that at least one mean Ra of the tested three groups of cooling condition is significantly different from the others.

4.3. Students' t-test on tool nose wear and Ra groups of AISI P20
Students' t-test on tool nose wear in each cooling condition resulted that, tool nose wear with dry cutting, and flood cooling is significantly lower than that of MQL. Compared to flood cooling, dry cutting has given significantly higher tool nose wear. t-test on Ra of P20 groups showed that, Ra with dry cutting, and flood cooling was significantly lower than that of MQL. Flood cooling has given significantly lower Ra than dry cutting. Results are presented in Table 5.

Table 5. Students' t-test results of tool nose wear, and Ra with P20.

	Tool nose wear		Ra	
	Flood	MQL	Flood	MQL
Dry	Significantly high	Significantly low	Significantly high	Significantly low
Flood		Significantly low		Significantly low

4.4. Students' t-test on tool nose wear groups of AISI D2
Students' t-test on tool nose wear in each cooling condition resulted that, tool nose wear with dry cutting, and flood cooling is significantly lower than that of MQL. Compared to flood cooling, dry cutting has given significantly higher tool nose wear. t-test on Ra of P20 groups showed that, Ra with dry cutting was significantly higher than that of MQL. However, flood cooling has given significantly lower Ra than both dry cutting, as well as lower than MQL. Therefore, Ra with MQL was lower than dry cutting, but higher than flood cooling. Results are presented in Table 6.

Table 6. Students' t-test results of tool nose wear, and Ra with D2.

	Tool nose wear		Ra	
	Flood	MQL	Flood	MQL
Dry	Significantly high	Significantly low	Significantly high	Significantly high
Flood		Significantly low		Significantly low

4.5. Effect of work-piece material on tool nose wear

Students' t-test on groups of tool nose wear of each material with each cooling condition shown in Table 7, proved that, there is no significant difference in mean tool nose wear between the two materials in each cooling condition. In dry cutting condition, the Ra of D2 was seen significantly higher than that of P20. However, in flood cooling, and in MQL, the Ra of D2 was seen to significantly lower that of P20.

Table 7. Comparison of tool nose wear, and Ra in each material.

	Tool nose wear			Ra		
	Dry (P20)	Flood (P20)	MQL (P20)	Dry (P20)	Flood (P20)	MQL (P20)
Dry (D2)	Not sig.			Sig'tly high		
Flood (D2)		Not sig.			Sig'tly low	
MQL (D2)			Not sig.			Sig'tly low

Variation of the mean tool nose wear, and mean Ra with respect to the three cooling methods, and two work material is shown in Figure 4. The differences in mean tool nose wear between the two materials was found to be not significant. D2 showed a larger reduction in Ra with flood cooling compared to dry cutting. D2 had a 15% surface roughness reduction in flood cooling compared to dry cutting, while P20 had only 7% reduction. However, the increase in Ra with MQL compared to flood cooling was, 13% in P20, and 11% in D2. P20 had 206% increase in tool nose wear in MQL compared to flood cooling, while D2 had 203% increment. Tool nose wear was reduced by 56% in P20 with flood cooling compared to dry cutting, while D2 showed a reduction of 46%.

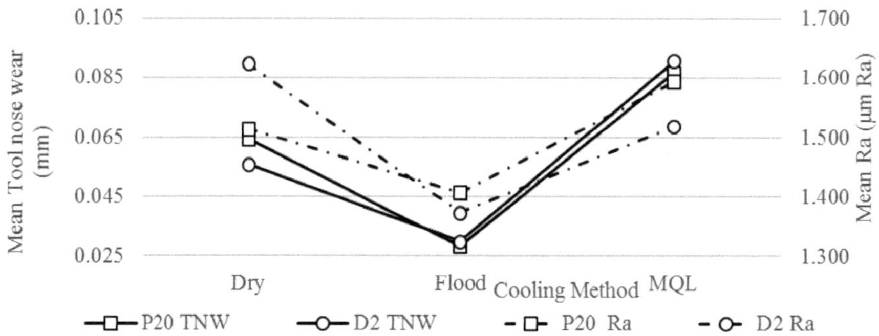

Figure 4. Mean tool nose wear (TNW), and mean Ra variation with cooling method, and work material.

5. Discussion

Although there is literature that suggests, MQL improves tool life and surface roughness in machining, this study found results that contradicts with them. However, it has seen by other researchers that MQL aerosol application has reduced tool life compared to flood cooling in some work-tool material combinations in certain machining conditions [19], [20]. It has been argued by many researchers that the performances of a cooling method, or machining performances in general, does depend on the work and tool materials. Evidence from this study supports that argument, as there is no similar patterns of tool nose wear or Ra with respect to the cooling methods.

Initially it was planned to take tool nose radius as the indicative measurement for tool wear. Radius of the tool nose after the tool was used for turning 500 mm of work-piece, was measured using the CMM. Upon usage, a reduction in the tool nose radius was expected but there were odd results which gave higher tool nose radius than the initial. Further investigation on this matter reveals that the curve approximation done for the tool tip in CMM leads to errors. Picking points on the worn edge with the CMM was vague and hence it was leading in to erroneous results.

The experiment was conducted with only two commonly used work-piece material and one type of tool. This has reduced the generalizability of the findings of the study. Therefore, the experiment needs to be extended to cover more work-piece material types, cutting tool materials, cutting fluid types and environmental conditions.

6. Conclusion

MQL aerosol cooling causes higher tool nose wear in machining AISI P20 and D2 material using coated carbide tools, compared to machining with conventional flood cooling method or dry cutting method.

The surface roughness of the work-pieces of both materials machined with MQL was higher than that obtained with flood cooling method. Particularly in P20 work-pieces, the surface roughness with MQL was even significantly higher to that of dry cutting as well.

Therefore, in terms of improving tool life, or surface finish of the machine parts, MQL does not provide a favourable performance with neither P20 nor D2 steels.

Tool nose wear in coated carbide tools, in machining AISI P20 and D2, in dry cutting, flood cooling condition, or MQL, is not significantly different. However, surface roughness varied significantly with the material. In dry cutting, surface roughness in D2 was significantly lower than P20, while the opposite occurred in flood cooling and MQL.

7. References

[1] E Brinksmeier, M Garbrecht, C Heinzel, T. Koch, and J Eckebrecht, "Current Approaches in Design and Supply of Metalworking Fluids," *Tribol. Lubr. Technol.*, pp. 36–49, 2009

[2] T Brockhoff and A Walter, "Fluid Minimization in Cutting and Grinding," *Abrasives*, pp. 38–42, 1998

[3] S Kalpakjian and S R Schmid, *Manufacturing Engineering and Technology*. Prentice Hall, 2010

[4] N Boubekri, "An Investigation in Drilling 1020 Steel Using Minimum Quantity Lubrication," *Int. J. Appl. Sci. Technol. Vol.*, vol. 1, no. 5, pp. 60–67, 2011

[5] S W M A I Senevirathne and M A R V Fernando, "Effect of Cryogenic Cooling on Machining Performance on Hard to Cut Metals - A Literature Review," in *National Engineering Conference 2012, 18TH ERU Symposium*, 2012, pp. 38–46

[6] H A A Kishawy, M Dumitrescu, E -G Ng, and M A A Elbestawi, "Effect of coolant strategy on tool performance, chip morphology and surface quality during high-speed machining of A356 aluminum alloy," *Int. J. Mach. Tools Manuf.*, vol. 45, no. 2, pp. 219–227, Feb. 2005

[7] Y Sun, B Huang, D A Puleo, and I S Jawahir, "Enhanced machinability of Ti-5553 alloy from cryogenic machining: Comparison with MQL and flood-cooled machining and modeling," *Procedia CIRP*, vol. 31, pp. 477–482, 2015

[8] M Z A Yazid, C H Che Haron, J A Ghani, G A Ibrahim, and A Y M Said, "Surface integrity of Inconel 718 when finish turning with PVD coated carbide tool under MQL," *Procedia Eng.*, vol. 19, pp. 396–401, Jan. 2011

[9] K Weinert, I Inasaki, J W W Sutherland, and T Wakabayashi, "Dry Machining and Minimum Quantity Lubrication," *CIRP Ann. - Manuf. Technol.*, vol. 1, no. 1, pp. 511–537, Jan. 2004

[10] M M A Khan, M A H Mithu, and N R Dhar, "Effects of minimum quantity lubrication on turning AISI 9310 alloy steel using vegetable oil-based cutting fluid," *J. Mater. Process. Technol.*, vol. 209, no. 15–16, pp. 5573–5583, Aug. 2009

[11] P W Marksberry and I S Jawahir, "A comprehensive tool-wear/tool-life performance model in the evaluation of NDM (near dry machining) for sustainable manufacturing," *Int. J. Mach. Tools Manuf.*, vol. 48, no. 7–8, pp. 878–886, 2008

[12] R Autret and S Y Liang, "Minimum Quantity Lubrication in Finish Hard Turning," in *HNICEM*, 2003, pp. 1–9

[13] S S Bhambale, C V Patil, and S R Nirmal, "Experimental Investigation on Wet Lubrication System in Turning Operation of Mild Steel by Minimizing the Flow Rate of Cutting Lubricants," *Int. J. Eng. Innov. Technol.*, vol. 1, no. 4, pp. 201–205, 2012

[14] N R Dhar, M T Ahmed, and S Islam, "An experimental investigation on effect of minimum quantity lubrication in machining AISI 1040 steel," *Int. J. Mach. Tools Manuf.*, vol. 47, no. 5, pp. 748–753, Apr. 2007
[15] A Attanasio, M Gelfi, C Giardini, and C Remino, "Minimal quantity lubrication in turning: Effect on tool wear," *Wear*, vol. 260, no. 3, pp. 333–338, Feb. 2006
[16] V Upadhyay, P K Jain, and N K Mehta, "Minimum Quantity Lubrication Assisted Turning - An Overview," in *Daaam International Scientific Book 2012*, K. BRANKO, Ed. Vienna: DAAAM International Vienna, 2012, pp. 463–478
[17] T Walker, *The MQL Handbook*. Unist Inc., 2013
[18] D C Montgomery, *Design and Analysis of Experiments Eighth Edition*, Eighth., vol. 2. John Wiley & Sons, Inc., 2013
[19] A D Jayal and A K Balaji, "Effects of cutting fluid application on tool wear in machining : Interactions with tool-coatings and tool surface features," vol. 267, pp. 1723–1730, 2009
[20] Y Kamata and T Obikawa, "High speed MQL finish-turning of Inconel 718 with different coated tools," *J. Mater. Process. Technol.*, vol. 192–193, pp. 281–286, Oct. 2007

Acknowledgements
The authors wish to thank the Senate Research Committee (SRC) of University of Moratuwa, Sri Lanka, for funding the research, and National Science Foundation (NSF) of Sri Lanka, for funding dissemination of research findings. Authors express their sincere gratitude for Mr. I.M.J Priyankara, and Mrs. N.H.S. Dias for their technical assistance.

IOP Conf. Series: Materials Science and Engineering **244** (2017) 012007 doi:10.1088/1757-899X/244/1/012007

Effect of commercial SiO_2 and SiO_2 from rice husk ash loading on biodegradation of Poly (lactic acid) and crosslinked Poly (lactic acid)

C Prapruddivongs[1], M Apichartsitporn[1], T Wongpreedee[2]

1Faculty of Science, Energy and Environment, King Mongkut's University of Technology North Bangkok, Rayong Campus, Rayong 21120, Thailand
2Faculty of Engineering and Technology, King Mongkut's University of Technology North Bangkok, Rayong Campus, Rayong 21120, Thailand
E-mail address: chana.p@sciee.kmutnb.ac.th

Abstract. In this work, biodegradation behavior of poly (lactic acid) (PLA) and crosslinked PLA filled with two types of SiO_2, precipitated SiO_2 (commercial SiO_2) and SiO_2 from rice husk ash, were studied. Rice husks were first treated with 2 molar hydrochloric acid (HCl) to produce high purity SiO_2 , before burnt in a furnace at 800ºC for 6 hours. All components were melted bending by an internal mixer then hot pressed using compression molder to form tested specimens. FTIR spectra of SiO_2 and PLA samples were investigated. The results showed the lack of silanol group (Si-OH) of rice husk ash after steric acid surface modification, while the addition of particles can affect the crosslinking of the PLA. For biodegradation test by evaluating total amount of carbon dioxide (CO_2) evolved during 60 days incubation at a controlled temperature of 58±2°C, the results showed that the biodegradation of crosslinked PLA occurred slower than the neat PLA. However, SiO_2 incorporation enhanced the degree of biodegradation In particular, introducing commercial SiO_2 in PLA and crosslinked PLA tended to clearly increase the degree of biodegradation as a consequence of the more accelerated hydrolysis degradation.

1. Introduction

Biodegradable polymers, such as poly (lactic acid) (PLA) [1-2], polyhydroxybutyrate (PHB) [3-4], cellulose [5-6] have been becoming increasingly popular. Among these, PLA is the most attractive polymers for further improvement because of its high mechanical strength, biocompatibility, biodegradability and processability facilitating the PLA to be used for much applicability [7-8]. However, the low thermal stability restricted the PLA in some applications [9]. In order to overcome the drawback, many technics, for example, creating crosslinking structures [10], forming composites [11-12] and chain extension [13] were employed. The electron beam irradiation associated with triallyl isocyanurate (TAIC) was normally utilized to form PLA crosslinking structures, resulting in an improvement of thermal stability of the PLA [14]. The inclusion of chemical crosslinking agent either TAIC and DCP at small dosage can also increase gel fraction and crosslinking density of the PLA [10]. Incorporation of SiO_2 1.5 wt% improved thermal stability of the PLA/SiO_2 composites approximately 20 °C as a result of a decrease in spherulite dimension causing an increase of crystallinity [15]. Enhancement of the thermal property and crosslinking structure can affect the biodegradation rate of the PLA. Thus, this work aimed to study the effect of SiO_2 loading, commercial SiO_2 and SiO_2 from rice husk ash, on the biodegradation of PLA and crosslinked PLA (CrPLA).

2. Experimental

2.1. Materials and chemicals
Poly(lactic acid) (PLA, 2003D, by NatureWorks, USA) was used in this work. Triallyl isocyanurate (TAIC) and dicumyl peroxide (DCP) were supply by PI industry Co. (Bangkok, Thailand). Synthetic precipitated silica was purchased from OSC Siam Silica (Bangkok, Thailand). Stearic acid (B18010 by, Kuala Lumpur Kepong Bhd, Malaysia) was used to modify surface of silica.

2.2. Silica preparation
Rice husks 100 g were cleaned by washing with tap water 3 times before soaking in 1 L of 2 M HCl solution for 4 hours. The husks were dried at 100°C for overnight in an oven after washing with tap water until pH was 7. Next, the dried husks were burnt at 800 °C for 6 hours. The rice husk ashes ($RHSiO_2$) and commercial silica powder ($CSiO_2$) were milled for 12 hours and sieved to select the SiO_2 size under 75 µm. Finally, silica were modified surface using stearic acid soluted in isopropyl alcohol. The ratio of stearic acid and silica was 1:10. After the surface treatment, SiO_2 ($CSiO_2$ and $RHSiO_2$) were dried 12 hours before mixing with PLA.

2.3. Specimen preparation
PLA was dried at 80°C for 3 hours to remove moistures. All components including PLA, $RHSiO_2$, $CSiO_2$, TAIC and DCP were-melt bending using an internal mixer (MX500-TQ, Chareontut, Thailand) to produce PLA and crosslinked PLA composites. The ratio of chemical crosslinking agent TAIC and DCP used was 0.5:0.5 weight percent. Testing specimens were formed by compression molded at 190 °C for 5 min.

2.4. Biodegradation test
The set apparatus for biodegradation test was applied from ASTM D5338 to determine the degree of biodegradations. In order to maintain thermophilic condition, temperature was controlled at 58±2°C throughout 60 days of the test. Carbon dioxide (CO_2) occurred were reacted with Barium hydroxide ($Ba(OH)_2$) solution. Then, the degree of biodegradation was defined by titration with hydrochloric (HCl). The schematic apparatus is shown in Figure 1.

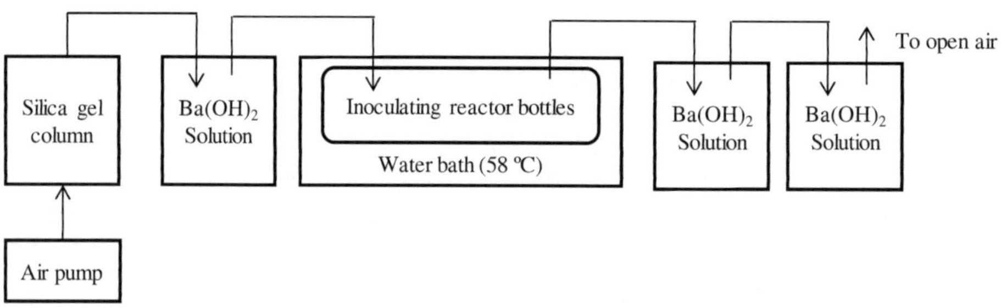

Figure 1. Schematic apparatus for aerobic biodegradation test

3. Characterizations
An attenuated total reflectance mode Fourier transform infrared spectroscopy (FTIR; Spectrum Spotlight 300, PerkinElmer, USA) was performed to examine surface treatment of SiO_2 and chemical crosslinking of PLA composites filled TAIC and DCP.

IOP Conf. Series: Materials Science and Engineering **244** (2017) 012007 doi:10.1088/1757-899X/244/1/012007

4. Result and discussion

FTIR of $CSiO_2$ and $RHSiO_2$ before and after surface treatment are given in Figure 2. FTIR spectrum of pure $CSiO_2$ appeared at about 3500 cm^{-1}, 1100 cm^{-1} and 950 cm^{-1} which belong to hydroxyl (-OH), siloxane (Si-O-Si) and silanol (Si-OH), respectively. Spectrum of pure $RHSiO_2$ was similar to the spectrum of $CSiO_2$. After stearic acid surface treatment, silanol group at about 950 cm^{-1} for $CSiO_2$ still remained. However, the silanol peak disappeared for treated $RHSiO_2$ indicating the reaction between carboxylic group (-COOH) and hydroxyl group (-OH) performed well. This disappearance confirms the accomplished coat of stearic acid on $RHSiO_2$ surface. Figure 3 shows FTIR spectra of pure PLA and chemical crosslinked PLA (CrPLA) with the addition of 4 wt% treated SiO_2. The carbonyl (C=O) stretching peak of PLA happened at the wavenumber of 1750 cm^{-1}. The incorporation of chemical crosslinking agents appeared a new peak at about 1690 cm^{-1} which clearly insists the crosslinking of TAIC and PLA [16]. However, the peek became unclear for $4RHSiO_2CrPLA$ and $4CSiO_2CrPLA$. This suggested a decrease in crosslinking level of CrPLA in the presence of $RHSiO_2$ and $CSiO_2$.

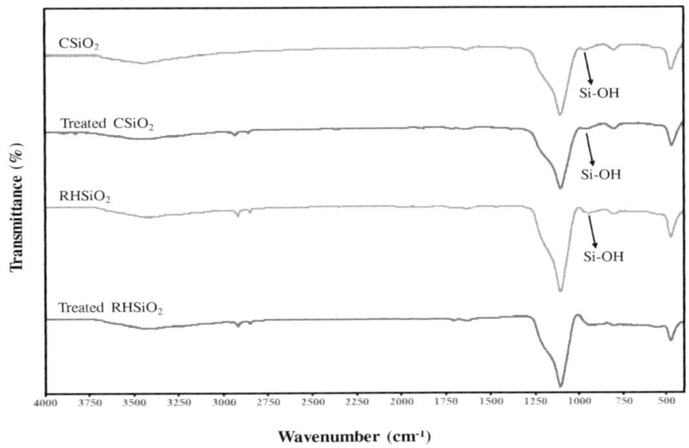

Figure 2. FTIR spectra of pure $CSiO_2$, $RHSiO_2$, treated $CSiO_2$ and treated $RHSiO_2$

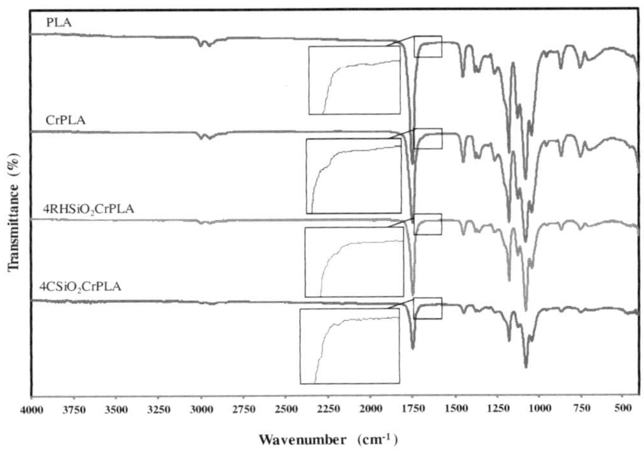

Figure 3. FTIR spectra of PLA, CrPLA , $4RHSiO_2CrPLA$ and $4CSiO_2CrPLA$

The degrees of biodegradation of PLA and crosslinked PLA composites are shown in Figure 4. It can be seen that all tested samples biodegraded increasingly upon the incubation time. The first 20 days of the test, the sample degraded slowly. This would imply that the long chain and high molecular weight of PLA could prevent water molecules to penetrate retarding hydrolysis reaction to occur. For later phase (21-60 days), biodegradation began rapidly. The more water molecules could penetrate; the more hydrolysis reaction would occur. The crosslinked samples slowed down the degree of biodegradation. The crosslinking reaction created dense PLA chains retraining the PLA to be hydrolyzed. It can be observed that added SiO_2 in PLA and CrPLA displayed enhancing degree of biodegradation. This effect was more remarkable for $CSiO_2$. After 60 days of testing, the degree of biodegradation of PLA, $1RHSiO_2PLA$ and $1CSiO_2PLA$ were 56.8%, 58.8% and 70.1%, respectively. The FTIR of SiO_2 surface modification (Figure 2) displayed successful surface treatment of stearic acid onto $RHSiO_2$ surface causing less hydrophilicity of $RHSiO_2$. The hydrophilicity played a key role in facilitation of the PLA as well as crosslinked PLA to be hydrolyzed.

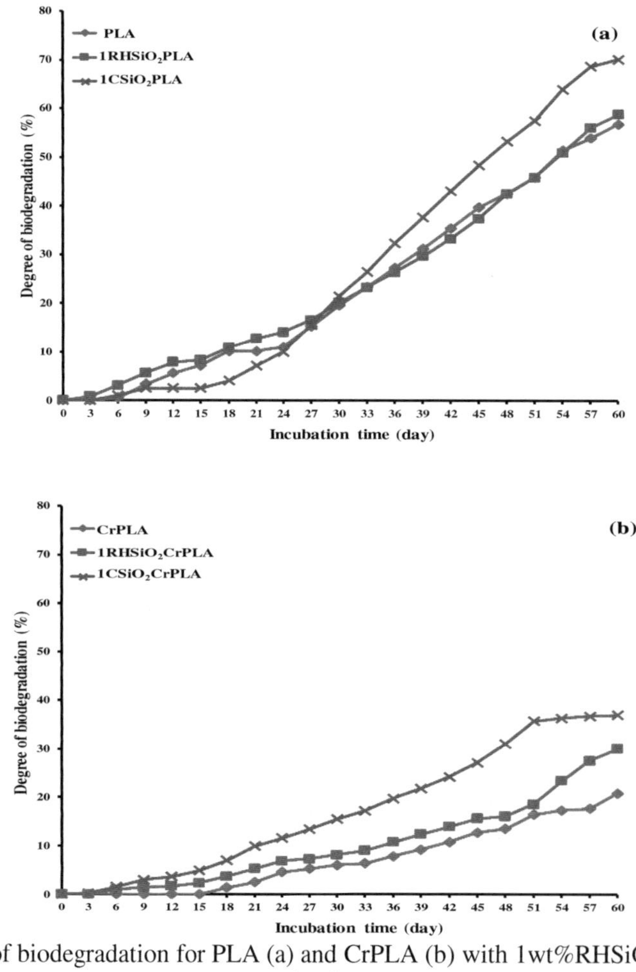

Figure 4. Degree of biodegradation for PLA (a) and CrPLA (b) with 1wt%$RHSiO_2$ and 1wt%$CSiO_2$ loading.

5. Conclusion

In this work, biodegradation behavior of poly (lactic acid) (PLA) and crosslinked PLA filled with precipitated SiO_2 (commercial SiO_2) and SiO_2 from rice husk ash, were studied. FTIR spectra of SiO_2 showed the lack of silanol group (Si-OH) of $RHSiO_2$ after steric acid surface modification suggesting the accomplished coat of stearic acid on $RHSiO_2$ surface. The addition of SiO_2 particles affected the crosslinking of the PLA. Biodegradation of crosslinked PLA occurred slower than the neat PLA. However, SiO_2 incorporation enhanced the degree of the degradation. In particular, introducing commercial SiO_2 in PLA and crosslinked PLA tended to clearly increase the degree and rate of biodegradation as a result of hydrophilicity.

6. Reference

[1] Jia WW, Gong RH, and Hogg PJ 2014 *Compos. Part. B.* **62** 104.
[2] Seligra PG, Nuevo F, Lamanna M, and Fama L 2013 *Compos. Part.* B. **46** 61.
[3] Melo JDD, Carvalho LFM, Medeiros AM, Souto CRO and Paskocimas CA 2012 *Compos. Part. B.* **43** 2827.
[4] Maiti P, Batt CA and Giannelis EP 2007 *Biomacromolecules.* **8** 3393.
[5] Fahmy TYA, Mobarak F, Fahmy Y, Fadl MH and El-Sakhawy M 2006 *Wood. Sci. Technol.* **40** 77.
[6] Fahmy TYA and Mobarak F 2008 *Carbohydr. Polym.* **72** 751.
[7] Lapruddivongs C and Sombatsompop N 2012 *Compos. B. Eng.* **43** 2730
[8] Prapruddivongs , C and Sombatsompop N 2015 *J. Thermoplast. Compos.* **8** 583
[9] Dorgan JR, Lehermeir H and Mang M 2000 *J. Polym. Environ.* **8** 1
[10] Yang SL, Wu ZH, Yang W and Yang MB 2008 *Polym. Test.* **27** 957
[11] Urayama H, Ma CH and Kimura Y, 2003 *Macromol Mater. Eng.* **288** 562.
[12] Trimaille T, Pichot C, Elaissari A, Fessi H, Briancon S and Delair T 2003 *Colloid Polym. Sci.* **281** 1184.
[13] Li BH and Yang MC, 2006 *Polym. Adv. Technol.* **17** 439.
[14] Jin FZ, Hyon SH, Iwata H and Tsutsum S 2002 *Macromol. Rapid Commun* **23** 909.
[15] Wu JH, Yen MS, Kuo MC and Chen BH 2013 *Mater. Chem. Phys.* **142** 726.
[16] Yang Sl, Wu ZH,Yang W and Yang MB 2008 *Polym. Test.* **27** 957.

Acknowledgement
The authors would like to thank King Mongkut's University of Technology North Bangkok (KMUTNB) for research funding. Contract no. KMUTNB-60-GOV-40.

Thermal properties of biopolyol from oil palm fruit fibre (OPFF) using solvolysis liquefaction technique

Shaharuddin Kormin[1], Anika Zafiah M. Rus[1*], M. Shafiq M. Azahari[1]

1Sustainable Polymer Engineering, Advanced Manufacturing and Material Center (SPEN-AMMC), Faculty of Mechanical and Manufacturing Engineering (FKMP), Universiti Tun Hussein Onn Malaysia (UTHM), 86400 Parit Raja, Batu Pahat, Johor, Malaysia
E-mail: shaharuddin_k@yahoo.com.my, *zafiah@uthm.edu.my

Abstract. Liquefaction is known to be an effective method for converting biomass into a biopolyol. The biomass liquefaction of oil palm fruit waste (PFW) in the presence of liquefaction solvent/polyhydric alcohol (PA): polyethylene glycol 400 (PEG400) using sulfuric acid as catalyst was studied. For all experiments, the liquefaction was conducted at 150°C and atmospheric pressure. The mass ratio of OPFW to liquefaction solvents used in all the experiments was, 1/3. Thermogravimetric analyses (TGA) were used to analyze their biopolyol and residue behaviors. It was found that thermal stability of oil palm mesocarp fibre (PM), oil palm shell (PS) and oil palm kernel (PK) fibre exhibited the first degradation of hard segment at (232, 104, 230°C) and the second degradation of soft segment at (314, 226, 412°C) as compared to PM, PS and PK residue which (229, 102, 227°C) of hard segment and (310, 219, 299°C) of segment, respectively. This behavior of thermal degradation of the hard segment and soft segment of biopolyol was changes after undergo solvolysis liquefaction process. The result analysis showed that the resulting biopolyol and its residue was suitable monomer for polyurethane (PU) synthesis for the production of PU foams.

1. Introduction

The oil palm is replanted every 25 to 30 years. Malaysia is one of the largest producer and manufacturer of palm oil products. 5 million tons of trunks ana 1.1 million tons of fronds are being disposed annually during replantation. Moreover, in the production process of palm oil and palm kernel oil in factories, 7 million tons of mesocarp, 4.1 million tons of shell, 1 million of palm kernel and 13.4 million tons of empty fruit bunches (EFB) are produced every years [1] [2]. This palm oil waste generates large amount of palm oil by-products, which can be recycled into biopolyol. If this biopolyol is put into good use, in this case as a main material in bio-polymer compound, then it will largely reduce the cost of high performance composite [2] [3]. At the same time, it will also reduce the amount of waste generated by the palm oil industry thus achieving a global aim of sustainable development. Not only it reduces the waste, it also preserves the nature by eliminating the need to harvest natural aggregates from natural sources.

Biopolyol has been widely used as a raw material for polyurethane production which has many applications in daily life such as foam synthesis, thermoset, thermoplastic and coating materials [4] [5]. As the limiting amount of polyol derived from petroleum and the demand of polyol is increasing, effort is needed to find out alternative raw materials in particular potential feedstock comes from agricultural waste.

However, there are limited studies focused on the use of oil palm fruit waste as material for synthesis of biopolyol. Oil palm fruit waste is very potential raw material for synthesis of biopolyol as it contains lignocellulose hence through solvolysis liquefaction process will produce biopolyol.
Thermogravimetric analysis (TGA) is an analytical technique that measures the weight of a small polymer sample as a function of time or temperature and hence gives a quantitative description of the thermal stability of material and the amount of the corresponding residue. The derivative weight loss curve known as differential thermogravimetric analysis (DTG) shown on the TGA themogram can be used to understand the onset decomposition temperature, the temperature at which the rate maximum of decomposition and decomposition temperature as well as the number of step involved in the thermal degradation of flexible polyurethane foam [5] [6]. TGA is useful to determine thermal stability and degradation behavior of a polymer or composite. The purpose of this research is to study the influence of independent variables (different types of biomass) on thermogravimetric analysis through liquefaction process of biopolyol.

2. Methodology

2.1. Materials

Materials used in this research were Oil palm fruit wastes (PW) which contain oil palm mesocarp fibre (PM), oil palm shell (PS) and oil palm kernel (PK) was obtained from Sindora Palm oil mills, Johor, Malaysia. Polyhydric alchohol (PA) such as polyethylene glycol 400 (PEG400) were used as liquefaction solvents and 98% percentage of sulfuric acid was used as the catalyst. All chemicals were purchased from Sigma-Aldrich.

2.2. Liquefaction of oil palm fruit fibre (OPFF) in PEG solution under conventional glass flask

Oil palm fruit fibre (OPFF) was liquefied at different liquefaction condition. Polyethylene glycol 400 (PEG400) was used as the main liquefaction solvent. First, oven dried oil palm fruit waste (20g) and liquefaction reagent were mixed 1:3 weight ratio with 5% sulfuric acid as a catalyst. Then the mixture was placed into a 250 mL three-branch flask equipped with thermometer and magnetic stirrer. Liquefaction temperature and time was 150°C and 120 min (2 hours), respectively. The liquefaction was carried out in an oil bath. After that, the liquefied oil palm wastes were dissolved in 100 mL of methanol for 4 h. The liquefied solutions were then vacuum-filtered through filter paper. The filtrated liquid was evaporated at 70°C using rotary evaporator to remove methanol, the obtained black liquid was designated crude biopolyols [7] [8]. Figure1 show the photograph of oil palm fruit fibre before and after solvolysis liquefaction process.

Samples	Oil Palm Fruit Fibre	Oil Palm Fruit Biopolyol	Oil Palm Fruit Residue
PM	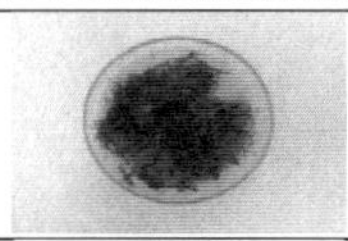		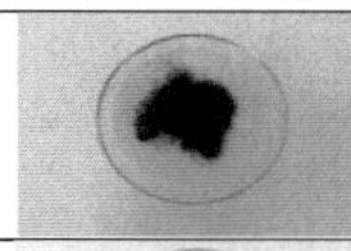
PS	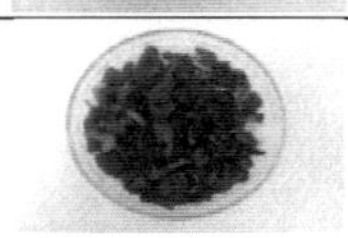	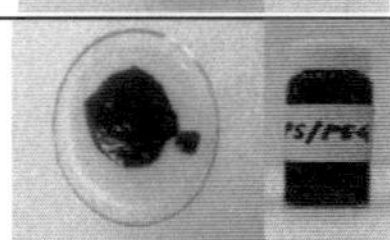	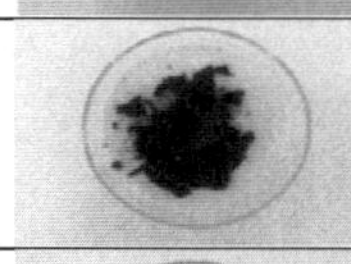
PK	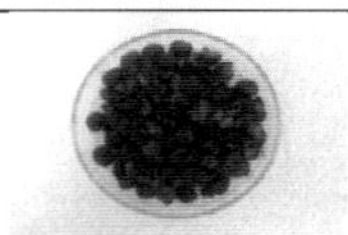	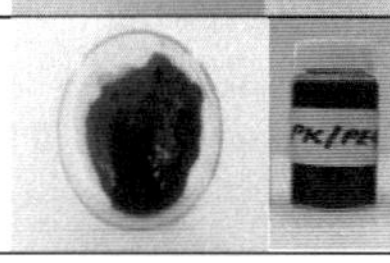	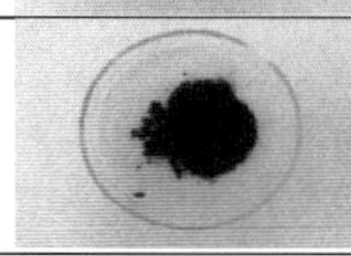

Figure 1. Photograph of oil palm fruit fibre before and after solvolysis liquefaction process.

2.3. Characteristics of polyol

Thermal Gravimetric Analysis (TGA)

TGA measurement was performed using *Linseis* TGA for characterized the thermal properties of biopolymer foam. The TGA for flexible and rigid biopolymer foam was measured weight loss and derivative weight loss curve was heated in the thermal gravimetric furnace from 20oC to 900oC at the heating rate of 10oC/min under oxygen atmosphere, flow rate 0.3 μL and using alumina crucible.

3. Result and discussion

Thermogravimetric analysis (TGA)

Figure 2 presents the thermogravimetric (TG) curves for oil palm fruite fibres (PFF) and its residue. The first point degradation correlates with the hard segment while the second peak correlates with the degradation of the soft segment. Qualitative characterization of the degradation process is elaborate by the onset and maximum peak temperature of the first step T_{1on} and T_{1max} along with the same for second step T_{2on} and T_{2max}. Detail TGA data, the onset decomposition temperature (T_{onset}) and the maximum decomposition temperature (T_{max}) for PFF and its residue are tabulated in Table 1.

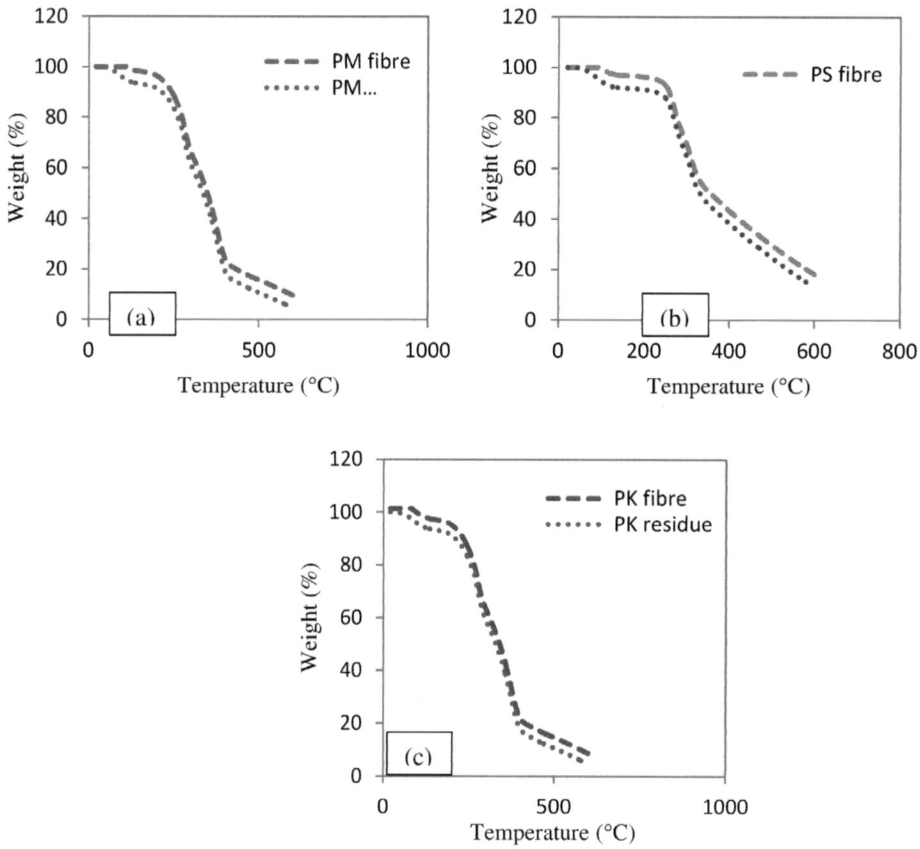

Figure 2. Thermogram weight loss of oil palm fruit fibre before and after liquefaction. (a) PM (b) PS (c) PK. Reaction condition: Oil palm fruit fibre/liquefaction solvent ratio, 1/3; temperature, 150°C; reaction time, 120 min; biomass loading, 20g; acid loading, 5%.

Table 1. Thermal decomposition of PFF and PFF residue.

Samples	PM fibre	PM residue	PS fibre	PS residue	PK fibre	PK residue
First Decomposition (°C)	232	229	104	102	230	227
T_{1on} (°C)	93	86	93	87	91	85
T_{1max} (°C)	290	287	210	207	288	284
Weight loss (%)	29	31	7	9	28	30
Second Decomposition (°C)	314	301	226	219	312	299
T_{2on} (°C)	290	288	210	207	288	284
T_{2max} (°C)	395	386	307	302	393	385
Weight loss (%)	51	53	39	41	49	51
Third Decomposition (°C)	410	402	311	305	409	399
T_{3on} (°C)	395	383	307	302	393	385
T_{3max} (°C)	596	594	596	594	593	592
Weight loss (%)	18	20	35	37	17	19

The weight loss of the first stage, which occurred from about 50 to 300°C, was due to the volatilization of methanol and water. It is clear that the decomposition of fibre characterized by a very small weight loss at around 100 C attributed to the release of residual water. The previous researchers was investigated the first peak of thermal degradation of biomass which is palm kernel shell, empty fruit brunch and palm mesocrap fiber due to moisture drying stage. The thermal decomposition observed below 200°C corresponded to the moisture evolution of the water linked to the hydroxyl and carbonyl groups of the fiber [6]. With increasing temperature during the TGA process, the decomposition lignin and other contents that are difficult to decompose caused the weight loss of fibres and residue at the second stage (300–400°C). the third stage of PFF and its residue largely attributed to the degradation of lignin and char residue from the second stage [9] [10]. When the TGA temperature reached the third stage of decomposition, the weight loss curve of biopolyols remained constant with a residual weight of approximately 10%, which consisted mainly of coke and ash. The dramatic difference between the TGA curves of fibre and its residue suggested that extensive reactions occurred among the components of fiber during the liquefaction process of biopolyols. In the process of liquefaction when using H_2SO_4 as a catalyst and PEG400, it was possible that the cellulose in the insoluble residues was changed into cellulose sulphate, which has the characteristic of thermal durability. Moreover, due to liquefaction, more active OH groups on PFF were possibly exposed in the polyols prepared with PEG400. Thus, the insoluble residues created from liquefaction process could act as a more efficient cross-linking agent [11].

From Figure 3(a), the DTG curve of PM fibre and PM residue degraded throughout the temperature range of 50–400°C and resulted in large mass losses, with one smaller peak at 100°C followed by a significantly large degradation maximum at 300°C. Figure 3(b) show the DTG graphs of PS fibre and PS residue with smaller peak at 98°C followed by two large maximum peak at 280°C and 350°C respectively. Figure 3(c) shows the DTG curve for PK fibre and PK residue. It was found that one smaller peak at 100°C and two large peak of degradation maximum at 250°C and 380°C. From the curves of differential thermogravimetry (DTG), it is evident that the decomposition mainly occurred in three successive stages above 50°C. The first peak appeared at the temperature range less than 100°C was attributed to the evaporation of moisture content and the beginning of weight loss of volatile components in the DTG evaluation profile of the samples [11] [12]. The thermal characteristic of PFF significantly changes after liquefaction process are shown in Figure 2 and Figure 3.

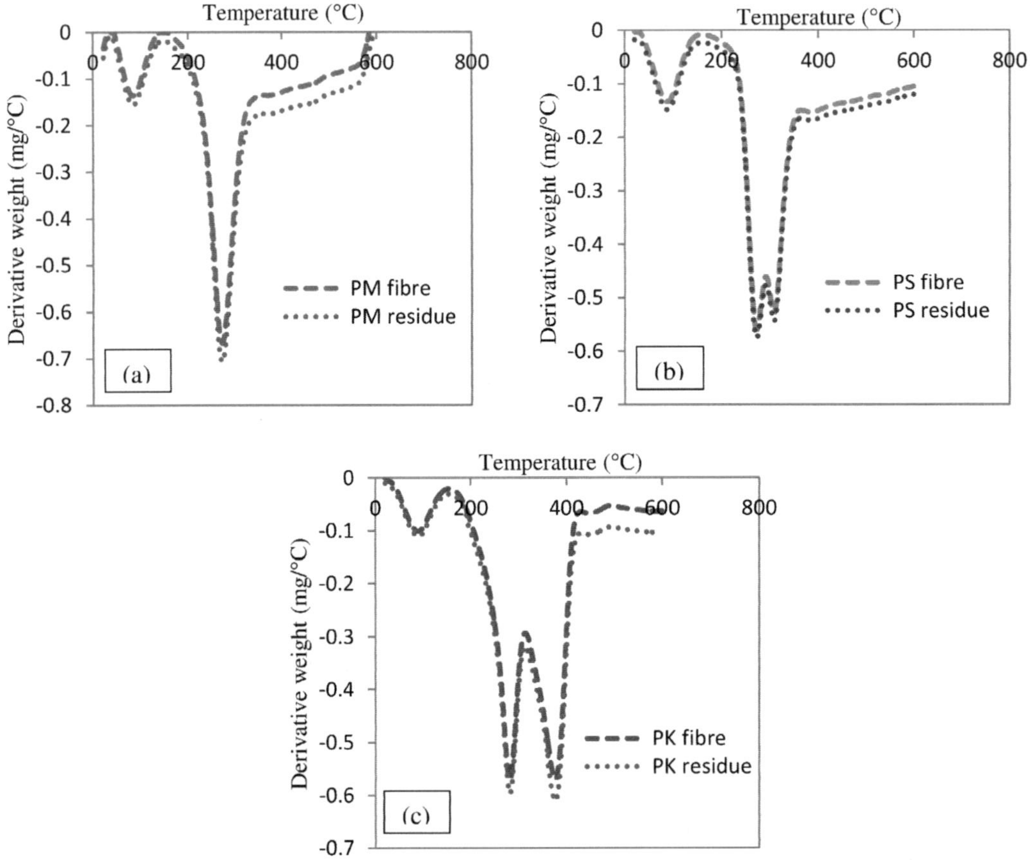

Figure 3. Thermogram DTG of oil palm fruit fibre before and after liquefaction. (a) PM (b) PS (c) PK. Reaction condition: Oil palm fruit fibre/liquefaction solvent ratio, 1/3; temperature, 150°C; reaction time, 120 min; biomass loading, 20g; acid loading, 5%.

4. Conclusion

The biopolyol based from oil palm fruits fibre (PFF) could be used as one of the raw material for preparation of polyurethane (PU) foam. The thermal degradation of PM fibre and residue exhibited the highest hard segment and soft segment as compared to PS fibre and PK fibre. The thermal degradation of oil palm fruit fibre is consistent after liquefaction process at the first degradation, second degradation and third degradation. This is due to thermal stability of samples at the hard segment has a great influence on the thermal stability of soft segment.

5. References

[1] S. Kormin and A. Z. M. Rus, "Preparation and Characterization of Biopolyol from Liquefied Oil Palm Fruit Waste : Part 2," vol. 400

[2] P. Rupani, M. Ibrahim, and S. Ismail, "Vermicomposting biotechnology: recycling of palm oil mill wastes into valuable products," *Int. J. Recycl. Org. Waste Agric.*, vol. 2, no. 1, p. 10, 2013

[3] S. H. Shuit, K. T. Tan, K. T. Lee, and A. H. Kamaruddin, "Oil palm biomass as a sustainable energy source: A Malaysian case study," *Energy*, vol. 34, no. 9, pp. 1225–1235, 2009
[4] E. Purwanto, J. Sutrisno, R. Apriliana, and K. M. Horax, "Research Article The optimization of biopolyol synthesis from liquefaction of rice straw using response surface method," vol. 7, pp. 46–53, 2015
[5] T. Gravimetric, A. Energy, A. Of, R. Polymer, and U. Ultraviolet, "Jurnal Teknologi THERMAL GRAVIMETRIC AND ACTIVATION ENERGY," vol. 32, pp. 57–64, 2015
[6] A. Z. M. Rus, N. Normunira, and M. Hassan, "Thermal Characteristic of Biopolymer Foam using Hot Compression Technique," no. November, pp. 10–11, 2014
[7] S. Kormin and A. Z. M. Rus, "Preparation and Characterization of Biopolyol From Liquefied Oil Palm Fruit Waste : Part 1," vol. 882, pp. 108–112, 2017
[8] S. Kormin and A. Z. M. Rus, "Preparation and Characterization of Biopolyol from Liquefied Oil Palm Fruit Waste : Part 2," vol. 882, pp. 113–118, 2017
[9] X. Luo, S. Hu, X. Zhang, and Y. Li, "Thermochemical conversion of crude glycerol to biopolyols for the production of polyurethane foams," *Bioresour. Technol.*, vol. 139, pp. 323–329, 2013
[10] J. D. Souza, "Polyurethane Foams Made from Liquefied Bark- Based Polyols Polyurethane Foams Made from Liquefied Bark-Based Polyols," no. August, 2014
[11] J. Xie, X. Zhai, C. Y. Hse, T. F. Shupe, and H. Pan, "Polyols from microwave liquefied bagasse and its application to rigid polyurethane foam," *Materials (Basel).*, vol. 8, no. 12, pp. 8496–8509, 2015
[12] T. Lee, Z. A. Zubir, F. M. Jamil, A. Matsumoto, and F. Y. Yeoh, "Combustion and pyrolysis of activated carbon fibre from oil palm empty fruit bunch fibre assisted through chemical activation with acid treatment," *J. Anal. Appl. Pyrolysis*, vol. 110, no. 1, pp. 408–418, 2014

Acknowledgement

The author would like to thanks Sustainable Polymer Engineering, Advanced manufacturing and materials Center (SPEN-AMMC), Universiti Tun Hussein Onn Malaysia (UTHM), Johor and Malaysian Government for supporting this research under FRGS vot 1481.

Acoustic properties of polymer foam composites blended with different percentage loadings of natural fiber

M. Shafiq M. Azahari[1], Anika Zafiah M. Rus[1*], Shaharuddin Kormin[1] and M. Taufiq Zaliran[1]

1 Sustainable Polymer Engineering, Advanced Manufacturing and Material Center (SPEN-AMMC), Faculty of Mechanical and Manufacturing Engineering, Universiti Tun Hussein Onn Malaysia, 86400 Parit Raja, Batu Pahat, Johor, Malaysia.

E-mail: muhammad.shafiq.azahari@gmail.com[1], zafiah@uthm.edu.my[1]*

Abstract. This study investigates the acoustic properties of polymer foam composites (FC) filled with natural fiber. The FC were produced based on crosslinking of polyol, with flexible isocyanates and wood filler. The percentages of wood filler loading are 10, 15, and 20 wt% ratio of polyol. The FC also has a thickness of 10, 20 and 30 mm. The acoustic properties of the FC were determined by using Impedance Tube test, Optical Microscope (OM) and Mettler Toledo Density Kit test. The results revealed that FC20 with 30 mm in thickness gives the highest sound absorption coefficient (α) with 0.970 and 0.999, at low and high frequency respectively. FC20 also shows smallest pores structures size with 134.86 μm and biggest density with 868.5 kg/m^3 which helps in absorbing sound. In this study, FC with different percentage loading of wood filler and different foam thickness shows the ability to contribute the absorption coefficient of polymeric foam at different frequency levels. Lastly, this type of FC is suitable for any type of sound absorption applications material.

1. Introduction

As the undesirable noise has become hazardous, the demand for residential safety and better environment becomes a major requirement to societies. Several studies focusing on sound absorbing materials have been performed to find the solution for this problems. This material plays important roles in the engineering of acoustic such as the control of industrial noise control, room acoustics, automotive acoustics and studio acoustics [1, 2]. Sawmill and furniture manufacturing industry have increased their production rate due to high demand. The wood waste product from this industry can increase environmental issues, where the wood need to be disposed of through burning or throw away into the river that causes pollution [3, 4]. The wood waste product should be reuse or recycle to create a valuable product for future as well as reducing the environment pollution [5–7]. This practice also uses the concept sustainability of technology manufacturing that we have known the sustainability is our measurement the credibility of most all the manufacturing industry by today to illustrate the green environment for future generation [8].

Polymer is known as engineering material because it is useful for many application. Polymeric foam also one of possible material to use and fabricated with waste wood dust [9–11]. Thus, it could generate into useful application of sound absorbing material. Sound characteristic with different ratio of polymer composite could contribute to high sound absorption at lower absorption coefficient level. The combination of this two material will form a new composite which has better sound properties. The composite is known as polymer foam composite (FC). The FC were developed based on crosslinking of polyol, flexible isocyanates and wood filler. In this study, Red Meranti wood was used because of its excessive sources of the furniture manufacturing. The acoustic characteristic properties (sound absorption coefficient) of the composite were determined to analysis the property filler in polymer foam composites. Red Meranti wood is chosen because of their availability. Wood dust is lignocellulosic by product of sawmill industry that is available at low cost [3, 7, 12]. Sawdust has been produced excessively in sawmill and furniture industry so we can try to recycle to valuable product to increase their sustainability in this industry.

2. Methodology

2.1. Polymer foam composites (FC) preparation

The foam was prepared by once shot method. All ingredient were mixed together to produce the FC. The FC is prepared by mixing the polyol and isocyanate at a ratio of 1.0:0.5 respectively. The polyol was poured into the plastic container approximately 3.0g and the wood filler was mixed with a polyol (10%, 15% or 20% of polyol weight ratio). Then, the isocyanate was poured together into the mixture. The mixture is allowed to cure and expand at room temperature for 24 hours. After the FC were completely cured, the samples were cut by using a knife and cylindrical shape of Teflon mold with a diameter of 100 mm and 28 mm [3, 10, 11]. Table 1 shows the proportion ratio of polyol, isocyanates and wood filler for the FC.

Table 1. The proportion ratio of polymer foam composites (FC).

	Percentage loading (g)		
Sample	Polyol (g)	Isocyanates (g)	Woods filler (g)
PU Foam	30	15	0
FC10	30	15	2.25
FC15	30	15	6.75
FC20	30	15	9

2.2. Impedance tube test

The sound absorption coefficient for the samples were measured using the Impedance Tube test that consists of a adjustable filter, propagation tube, large sample tube 100 mm diameter for low frequency 0 to 1500 Hz, small sample tube 28 mm diameter for high frequency 1500 to 6000 Hz and two-microphone method and a digital frequency analysis system for the measurement of normal incidence sound absorption coefficient and normal specific acoustic impedance ratios of materials. Sound absorption of polymer foam composites was measured at 0-6000 Hz according to ASTM E1050.

2.3. Optical microscope (OM) test

OM was used to identify the effect of filler loading on the size of pore polymer foam composites (FC). This test was very important due to the filler loading affected the acoustic properties. The FC samples were prepared in small size in order to fit into the sample holder. The magnification of OM lens was adjusted into 10x magnification. The type of OM used is Nikon Eclipse LV150NL.

2.4. Mettler toledo density kit (HR250AZ) test

This density test was conducted to identify the difference in density of foam before and after added with the wood filler. Four samples for every polymer foam composites (FC) with size 10 x 10 x 10 mm were prepared to get the accurate value of density by using the average result. This research was used Buoyancy Method according to European Standard EN 993-1 to identify the density of FC. Equation (1) below is used to calculate the density.

$$\text{Density}, \rho = \frac{\text{Weight of sample in air (g)}}{|\text{Weight of sample in air (g)} - \text{Weight of sample in liquid (g)}|} \times \text{Density of liquid} \left(\frac{\text{g}}{\text{cm}^3}\right) \quad (1)$$

where, density of liquid (methanol) = 0.798 g/cm^3

3. Results and discussions

3.1. The effect of percentage loadings and thickness

Figure 1 graph shows the sound absorption coefficient level (α) results with three different thickness (10 mm, 20 mm and 30 mm) of polymer foam composites (FC) by using the Impedance Tube test. Figure 1(a) shows the result for PU foam, where, at low and high frequency region the highest α is contributed from 30 mm thickness foam with 0.772 and 0.960 respectively. Furthermore, Figure 1(b) shows the results for FC with 10% wood filler loading. At low and high absorption frequency level, the highest α is contributed from the highest thickness of 30 mm foam with 0.951 and 0.993 respectively. Figure 1(c) shows the result of FC for 15% percentage loading, where, at low and high frequency absorption level the highest α is 0.969 and 0.998 respectively with thickness 30 mm. Meanwhile, Figure 1(d) also shows the result of FC for 20% percentage loading which is at low and high frequency absorption level the highest α is contributed from 30 mm thickness foam with 0.970 and 0.999 respectively. In comparison, FC20 gives the highest α as compared to the overall results for PU foam and others FC.

The highest α is due to the highest FC thickness and wood filler percentage loading, the higher the sound resonant frequency will accumulate and it will dissipate energy of sound more efficiently through the foam [13]. The friction will increase when thickness increase due to the longer the path of the sound wave, and increase the sound absorption coefficient [11,14–16]. The highest absorption occurs at a resonant frequency of quarter wavelength of the wave sound [15]. A porous material is effective from the sound absorption point of view when the thickness is approximately one-tenth the wavelength of the incident sound [15–17]. Increase in sound absorption especially at low frequencies when the material thickness increases; at high frequencies, material thickness has a lower effect on sound absorption [15].

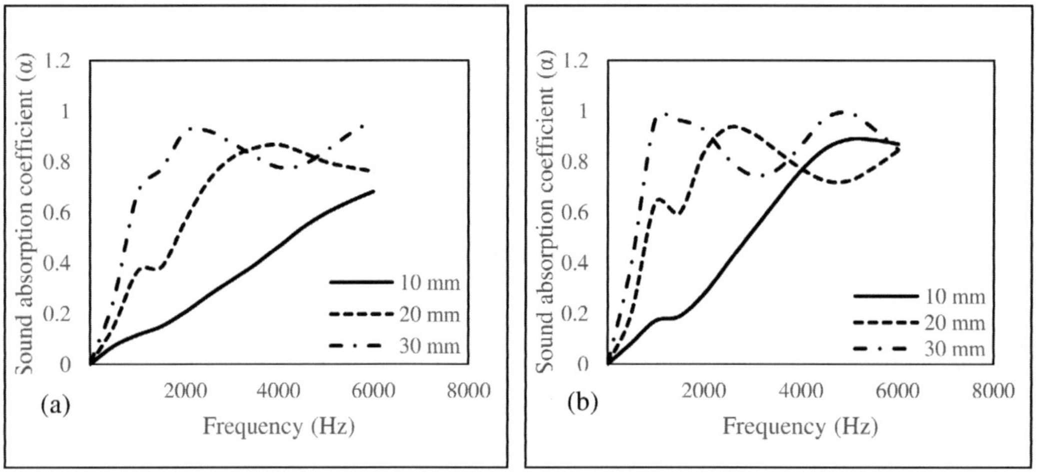

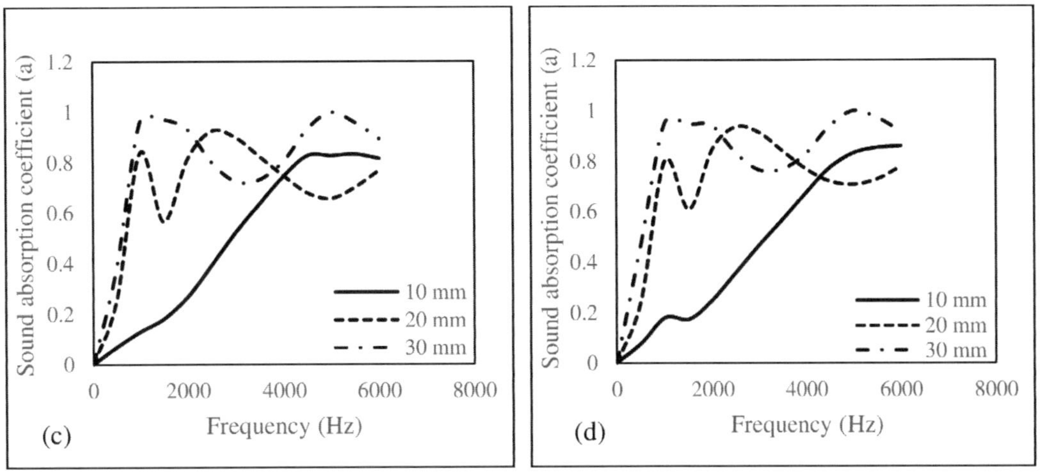

Figure 1. The sound absorption coefficient level (α) with different thickness of: (a) PU foam; (b) FC10; (c) FC15; and FC20.

3.2. The effect of pore size

Figure 2 shows the microstructure images of polymer foam composites (FC) and Figure 3 shows the average pore size (μm) results for overall FC and PU foam. The smallest average pore size for FC is FC20 followed by FC15, FC10 and PU foam with 134.86 μm, 168.82 μm, 175.22 μm and 185.21 μm respectively. Based on the average size of pore for overall samples, it is best to conclude that with the increasing of filler percentage loading, it will decrease the pores size of polymer form composite. The pores size allows the sound wave passing through the sound absorbing material and dissipated sound energy and damped itself [3, 4, 18]. Smaller pore size gives more collision occurs between the sound wave and cell wall, longer the path of reflection and refraction and more the absorption energy, thus, create greater sound absorption coefficient [3, 11].

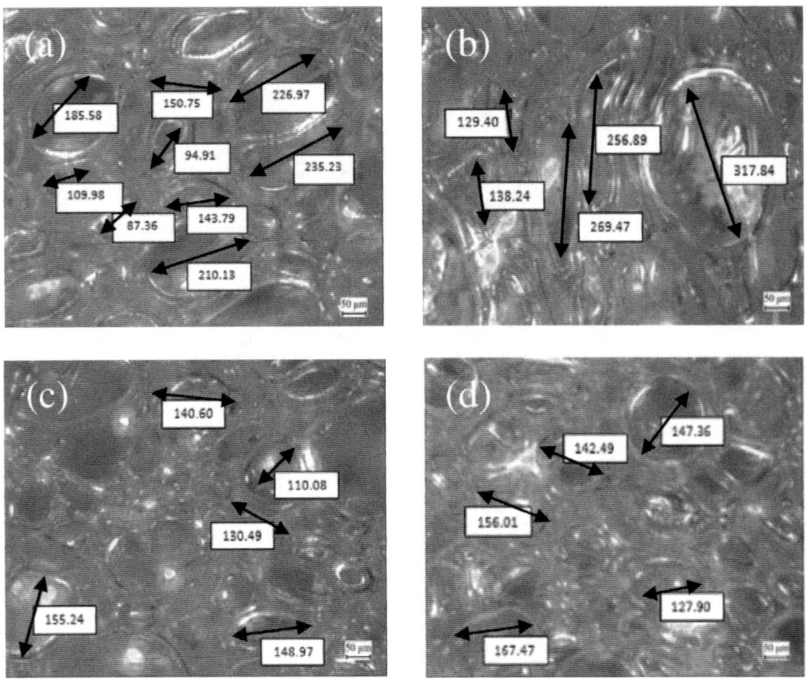

Figure 2. Microstructure image of (a) PU foam; (b) FC10; (c) FC15; and (d) FC20.

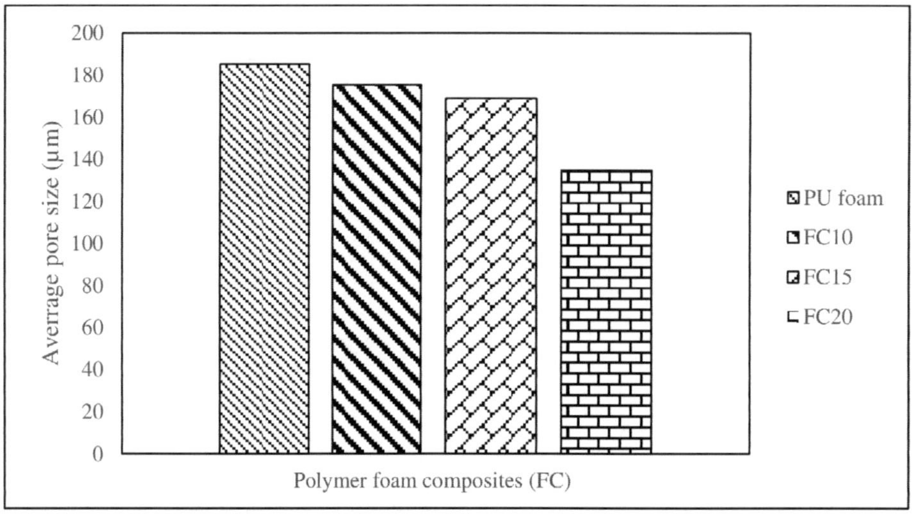

Figure 3. The average pore size (μm) for overall polymer foam composites.

3.3. The effect of density

Figure 4 shows the average value of density for four samples of polymer foam composites (FC) with different percentage loading of wood filler loading. It shows that the PU foam gives smallest density value of 827.4 kg/m^3. FC with 10%, 15% and 20% shows increment in density value of 844.0 kg/m^3, 850.8 kg/m^3 and 868.5 kg/m^3 respectively. FC with wood filler has a greater impact on the density values of the increase of percentage filler loading compared to PU foam. This result shows that the increase of density enhanced the ability to absorb more acoustic energies. This is because the number of fiber increases per unit area when the density is increasing and makes more sound energy losses as the surface friction increase, thus makes the sound absorption coefficient, α increases [15, 17, 19].

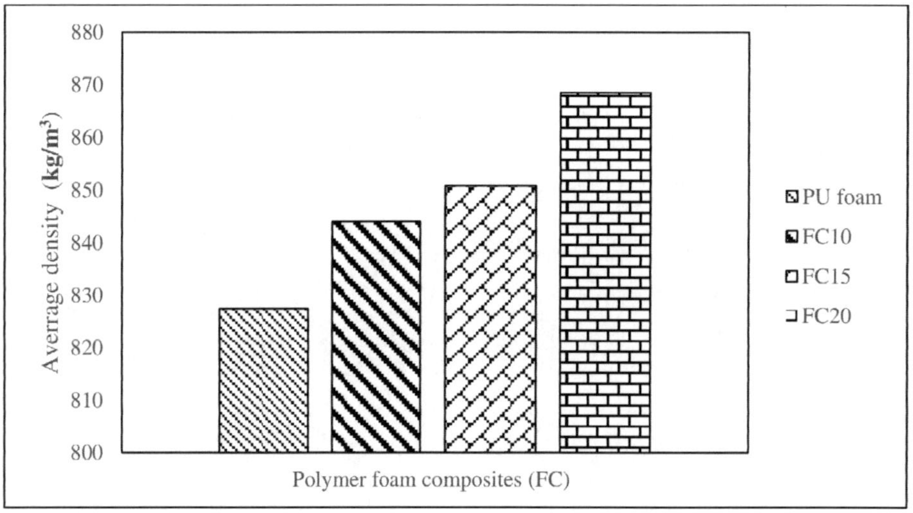

Figure 4. Average density (kg/m^3) for overall polymer foam composites

4. Conclusion

The results obtained concluded that the studies are successfully done. The analyzed data shows that acoustics properties of polymer foam composite are affected by different percentage loading of wood filler, thickness, pore size and density. FC20 with 30 mm in thickness gives the highest of the sound absorption coefficient, α with 0.970 and 0.999 at low and high frequency level respectively. FC20 also gives the smallest pore size with 134.86 μm and bigger density with 868.5 kg/m^3. In conclusion, the best sample material from this study it is from a wood virgin, that mostly gets the higher value of percentage wood filler loading and the higher thickness of foams to create the perfect material of sound absorption coefficient.

5. References

[1] Sung G, Kim J W and Kim J H 2016 Fabrication of polyurethane composite foams with magnesium hydroxide filler for improved sound absorption *J. Ind. Eng. Chem.* **44** 99–104
[2] Chathurangani O, Perera W and Kumari H 2012 Utilization of Sawdust and Coconut Coir Fibre as Noise Reducing Wall Surface Materials *Civil.Mrt.Ac.Lk* 16–9
[3] Rus A Z M, Shafizah S and Samples a P 2015 Acoustic Behavior of Polymer Foam Composite of Shorea leprosula after UV-Irradiation Exposure *Int. J. Mech. Aerospace, Ind. Mechatronics* **9** 188–92
[4] Sa'adon S and Rus A Z M 2014 Acoustical behavior of treated wood dust-filler for polymer foam composite *Appl. Mech. Mater.* **465-466** 1039–43
[5] Riduan M A, Yasmin Z and Harith H A Study on Dried Banana Stem Fiber and Wood Dust as Sound Absorbing Wall Panel **6** 89–100
[6] Zhao J, Wang X M, Chang J M, Yao Y and Cui Q 2010 Sound insulation property of wood-waste tire rubber composite *Compos. Sci. Technol.* **70** 2033–8
[7] Sa'adon S and Rus A Z M 2013 Utilization of Treated Red Meranti Wood Dust as Polymer Foam Composite for Acoustic Study *Key Eng. Mater.* **594-595** 760–4
[8] Putra A, Abdullah Y, Efendy H, Farid W M, Ayob M R and Py M S 2013 Utilizing sugarcane wasted fibers as a sustainable acoustic absorber *Procedia Engineering* vol 53 pp 632–8
[9] Hung T C, Huang J S, Wang Y W and Lin K Y 2014 Inorganic polymeric foam as a sound absorbing and insulating material *Constr. Build. Mater.* **50** 328–34
[10] Rus A Z M, Normunira N and Hassan M 2014 Thermal Characteristic of Biopolymer Foam using Hot Compression Technique 10–1
[11] Hassan N N M and Rus A Z M 2013 Influences of Thickness and Fabric for Sound Absorption of Biopolymer Composite *Appl. Mech. Mater.* **393** 102–7
[12] Jayamani E 2015 Sound Absorption and Impedance Study of Lignocellulosic Fibre based Composites for Acoustical Applications
[13] Zulkifli R, Zulkarnain and Nor M J M 2010 Noise control using coconut coir fiber sound absorber with porous layer backing and perforated panel *Am. J. Appl. Sci.* **7** 260–4
[14] Lou C W, Li T T, Huang C L, Hsu Y H and Lin J H 2015 Sound-Absorbing and Flame-Retarding Property of Nonwoven Compounded PU foam Planks *J. Eng. Fiber. Fabr.* **10** 29–38
[15] Seddeq H S 2009 Factors Influencing Acoustic Performance of Sound Absorptive Materials *Aust. J. Basic Appl. Sci.* **3** 4610–7
[16] Mcgrory M, Cirac D C, Gaussen O, Cabrera D and Engineers G 2012 Sound absorption coefficient measurement: Re-examining the relationship between impedance tube and reverberant room methods *Proc. Acoust. 2012 - Fremantle* 1–8
[17] Nordin M N A A, Wan L M, Zainulabidin M H and Kassim A S M 2016 RESEARCH FINDING IN NATURAL FIBERS SOUND ABSORBING MATERIAL **11** 8579–84
[18] Gayathri R, Vasanthakumari R and Padmanabhan C 2013 Sound absorption, thermal and mechanical behavior of polyurethane foam modified with nano silica, nano clay and crumb rubber fillers *Int. J. Sci. Eng. Res.* **4** 301–8
[19] Mahzan S, Zaidi a M A, Arsat N, Hatta M N M, Ghazali M I and Mohideen S R 2009 Study on Sound Absorption Properties of Coconut Coir Fibre Reinforced Composite with Added Recycled Rubber *Integr. Eng.* 1–6

Acknowledgement
The author would like to thank the supervisor and lab mates from SPEN-AMMC for their great help. The author also would like to thank Universiti Tun Hussein Onn Malaysia (UTHM), Johor, Malaysia and Malaysian Government for supporting this research under Fundamental Research Grant Scheme vot 1481.

A novel technique for producing conductive polyurethane nanofibrous membrane for flexible electronics applications

A Shaker[1], Ahmed H Hassanin[1,2], N M Shaalan[1], M A Hassan[1] and Ahmed Abd El-Moneim[1]

1 Materials Science and Engineering Department, Egypt-Japan University of Science and Technology, New Borg El-Arab, Alexandria 21934, Egypt
2 Textile Engineering Department, Faculty of Engineering, Alexandria University, Alexandria 21544, Egypt
E-mail: ahmed.abdel-baset@ejust.edu.eg

Abstract. A novel technique for producing stretchable and flexible conductive polyurethane nanofibrous membrane for flexible electronics applications has been developed. One of the most important challenges in fabricating polymeric membranes for flexible electronics is to enhance its sensitivity and conductivity maintaining its flexibility and stertchability. Wet electrospinning technique was used to fabricate thermoplastic polyurethane (TPU) nanofibers through a coagulant bath of a conductive grade co-polymer poly(3,4-ethylenedioxythiophene) poly(styrenesulfonate) (PEDOT: PSS). A Kapton sheet was used as a substrate to load TPU/PEDOT: PSS nanofibers on it. The morphology and stability of the as-spun TPU/PEDOT: PSS nanofibers has been investigated. The fabricated TPU/PEDOT: PSS nanofibrous membrane has showed a well stabilized ohmic behaviour under a wide range of different temperatures up to 90 ℃ as well as different values of tensile strain up to 4.43 %, which makes it suitable for high temperature strain sensing applications.

1. Introduction

Flexible and wearable electronic devices have become of a rapidly increasing demand in modern societies. [1-3] It is virtually impossible for conventional rigid electronics to achieve flexibility, stretchability, and deformability into different shapes. [4] There has been rapid progress on fabricating flexible or stretchable strain sensors. Strain sensors with high performance implicate many features, including high sensitivity, stability, stretchablity, response speed, process and production costs. [5-7] High-strain sensors got strong demand in many fields such as biomechanics, physiology and kinesiology applications areas [8, 9]. Recently, nanoscale materials were used to make novel strain sensors. For example, the device based on many typical structures with nanoparticles, nanowires, nanotubes, and graphene have been reported. [10-14]

Electrospinning is a versatile and simple fabricating method of fibers with diameters range of a few micrometers to ten nanometers or less. In conventional electrospinning process, a Taylor cone is used to expel a charged jet and rushes it under an electric field with a driving force onto the ground collector. After the evaporation of solvent, solid fibers with uniform diameter are randomly laid down on the collector [15]. In the past decade, diverse ultrathin fibers extracted from polymer, ceramic, metal, and glass have been synthesized by electrospinning, and their latent applications in optoelectronics, sensors, filters, fiber reinforcement, catalysis, drug delivery, supercapacitors, lithium batteries and tissue engineering have also been broadly investigated [16, 17].

Wet electrospinning is an altering technique of the common electrospinning technique. The modification is based on the usage of a liquid bath collector rather than a solid metallic one. Wet electrospinning was introduced for the first time as a method for producing nanofibrous scaffolds for tissue engineering at the laboratory scale. [18] It is a crafty matter to produce bulky and fluffy membrane using the common electrospinning technique. This task could be accomplished either using distinctive collectors [19] or by using porogen particles, that is, chemical blowing agents, settled in between the nanofibers [20]. Wet electrospinning is a relatively simple and effective method to produce three dimensional (sponge-like) materials without intricated devices and without especial chemical additives. [21]

High electrically conductive nanofibers have been produced from a range of organic conductors, such as polyaniline, polypyrrole, poly(3,4-ethylenedioxythiophene): poly(styrenesulfonate) (PEDOT: PSS), carbon nanotubes, and recently from graphene and graphene oxide. [22] In spite of its high conductivity, these fibers have significant poor deformability, having relatively limited to no elasticity or elastic recovery and cannot be stretched by more than 20% of their original lengths. On the other hand, fibers which are typically produced from elastomers, with very high range of elasticity and deformability are electrically insulating. The combination of high conductivity and high stretchability is considerable in applications requiring strain sensing such as stretchable circuits, wearable bionics and electrochromic textiles. [23]

In this work, stretchable conductive nanofibrous mate was developed with an enhanced range of measured electrical resistance at KΩ scale, where MΩ scale resistance is common for measurements of nanofibrous mats under normal circumstances. [24] The developed nanofibrous mate was produced by wet electrospinning technique where Thermoplastic Polyurethane (TPU) was electrospun into liquid bath which has PEDOT: PSS aqueous solution as a coagulant. Combining between enhanced electrical conductivity supported by PEDOT: PSS co-polymer solution and the stretchability of TPU elastomer in a mate at nanoscale with a stable ohmic behaviour makes it a good candidate for many applications such as supercapacitors, polymeric solar cells and strain gauge sensors.

2. Experimental details

2.1 Materials

Thermoplastic Polyurethane (TPU) C95 with a molecular weight of 107,020 g/mol and Polydispersity Index (PDI) of 1.83 was supplied by BASF Corporation, Germany. TPU renders various chemical structures, which enables it to be used for various applications like membranes, filtration, textiles and protective clothing. [25] Dimethylformamide (DMF 98%) was used as a solvent for TPU to prepare the polymer solution. Poly(3,4-ethylenedioxythiophene) poly (styrene sulfonate) (PEDOT: PSS) is a conductive grade co-polymer, which was supplied by Sigma-Aldrich, USA. This PEDOT: PSS is 1.3 wt.% of the colloidal dispersed polymer in distilled water. A typical Kapton sheet of thickness 75 μm and an area factor of 9.1 m2/kg supplied by UPILEX-Japan was used to be as a substrate for TPU nanofibers mate.

2.2 Developed wet electrospinning technique

TPU was dissolved in DMF at a concentration of 10 wt.% for five hours stirring at 60 ℃, then transferred into 5 ml syringe of electrospinning setup. NANON electrospinning setup was used to fabricate TPU nanofibers. During electrospinning process, a 30 kV direct current voltage was applied to TPU syringe and the collector over which the coagulant bath of PEDOT: PSS was set. A feed rate of 1 mL/h was used at a vertical distance of 10 cm between the needle tip and the collector bath. At the bottom of PEDOT: PSS collector bath, the Kapton substrate was placed. A membrane of average thickness of 60 μm was formed on the top surface of PEDOT: PSS solution which is instantaneously saturated by the conductive polymer solution. A schematic diagram of the wet electrospinning process is shown in Figure 1.

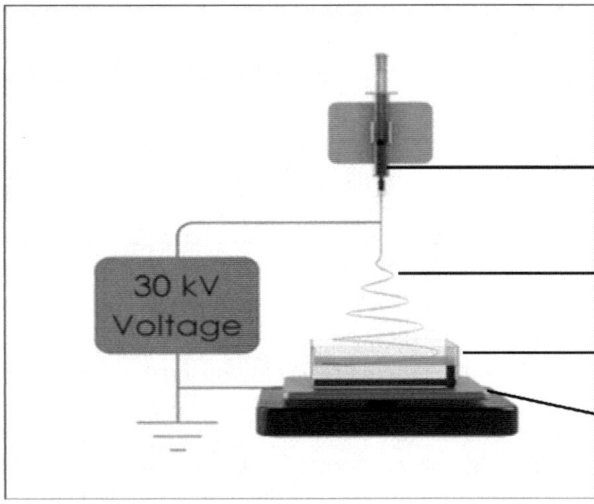

Figure 1. A schematic diagram of wet electrospinning technique.

2.3. Characterization of TPU/PEDOT: PSS nanofibrous membrane

Morphology of TPU/PEDOT: PSS nanofibrous membrane was investigated using JCM-6000PLUS NeoScope Benchtop scanning electron microscope (SEM). To study the I-V behaviour of the nanofibrous membrane, PHOTO EMISSION TECH. I-V test system was used. Fourier transform infrared (FTIR) spectrometer (Vertex 70, Bruker Scientific Instruments, Germany) and X-ray diffraction (XRD) spectrometer (LabX XRD-6100 Shimadzu, Japan) were used to analyse the TPU/PEDOT: PSS nanofibrous membrane.

3. Results and discussions

3.1. Morphology

Scanning electron microscope (SEM) was used to characterize the morphology of the as-spun TPU/PEDOT: PSS nanofibers. It was noticed that the nanofibers had smooth surface and appeared to be straight with no beads, as illustrated in Figure 2. The average diameter of the nanofibers is approximately 170 nm. It is obvious from morphology of nanofibrous mate that thin layers of PEDOT: PSS was formed between the nanofibers. This can be attributed to using PEDOT: PSS aqueous solution as a coagulant, where the phase inversion process is carried out. In this process, the exchange between the solvent [DMF] and non-solvent [coagulant, PEDOT: PSS aqueous solution] is occurred, which resulting in solidified TPU nanofibers and forming a thin layer of PEDOT: PSS covering the nanofibers and sometimes in between the nanofibers.

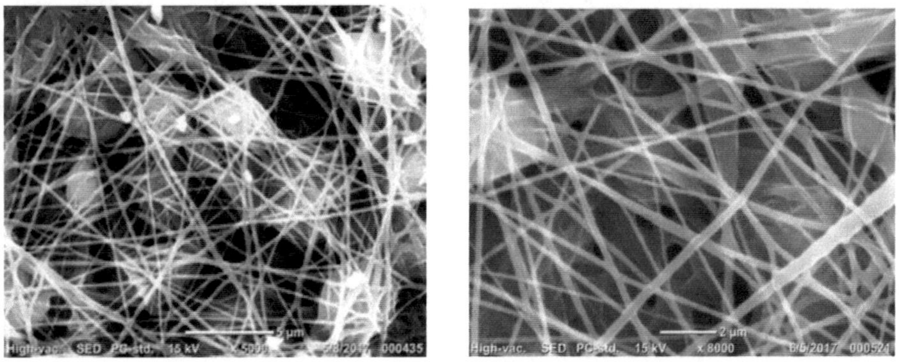

Figure 2. SEM images of TPU/PEDOT: PSS nanofibrous membrane.

3.2. XRD and FTIR analyses

The XRD pattern of TPU/PEDOT: PSS nanofibrous membrane is shown in Figure 3. TPU has diffraction peaks at 2.4°, 4.9°, 7.4° and 24.7°, and PEDOT: PSS has a diffraction peak at around 25° according to International Centre for Diffraction Data. Jia et al. [26] stated that if there is no interaction or only weak interaction between components in polyblend fibers, there would be two different crystalline peaks for each component. In this study, only one diffraction peak is seen for nanofibrous membrane around 19.0° due to the strong interaction between PEDOT: PSS and TPU. The diffraction peak is seen at 2Θ =19.0° for TPU/PEDOT: PSS nanofibers. The FTIR spectroscopy analyses for TPU, PEDOT: PSS and TPU/PEDOT: PSS nanofibers are shown in Figure 4. The spectra showed all the bonds of the chemical structure of TPU, PEDOT: PSS and TPU/PEDOT: PSS nanofiber composite.

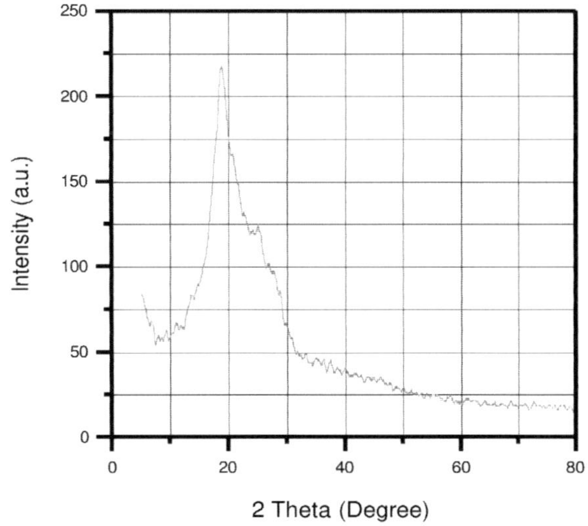

Figure 3. XRD pattern of TPU/PEDOT: PSS nanofibrous membrane.

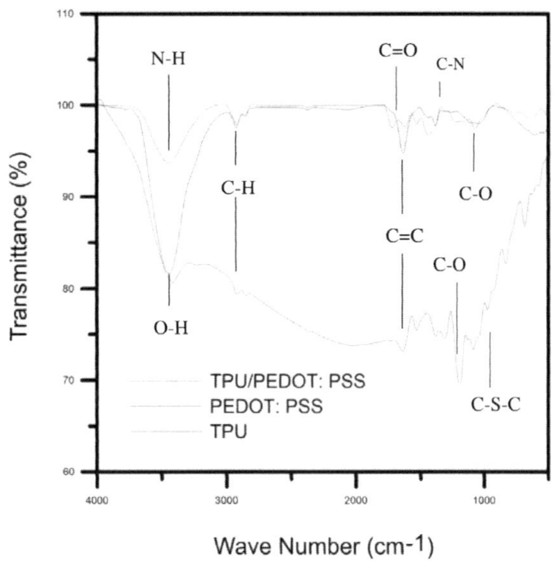

Figure 4. FTIR of TPU, PEDOT: PSS and TPU/PEDOT: PSS nanofibers.

3.3. I-V Characterization

Figure 5 shows I-V characterization curves of the as-spun TPU/PEDOT: PSS nanofibrous membrane. The I-V behaviour of produced nanofibrous membrane under different temperatures is illustrated in a semi-log plot as seen in Fig. 5(a) whereas Fig. 4(b) shows the I-V behaviour of the produced nanofibrous membrane under different strains in normal plot. From Fig. 5(a), it is clear that the membrane has showed a stabilized and symmetric ohmic behaviour at different values of temperature ranging from room temperature to 90 °C at zero strain. On the other hand, current-voltage characteristics of the TPU/PEDOT: PSS nanofibrous membrane have shown the normal linear trends under different percentages of tensile strain at room temperature. The produced mate, by the current developed technique, has demonstrated an ohmic behaviour irrespective the applied strain and the current decreases in a monotonic way with increasing the applied strain as shown in Fig. 4(b).
Based on the calculated conductivity obtained from these curves, Figs 4 (a) and (b), the as-spun TPU/PEDOT: PSS nanofibrous membrane has provided a great enhancement in the electrical conductivity compared to previous studies [24]. For instance, in this study, the TPU/PEDOT: PSS nanofibrous membrane has a KΩ resistance and the average value of the electrical conductivity was 0.08 S Cm^{-1}. However, according to other published work [24], a MΩ scale for electrical resistance of nanofibrous membranes is usually common under similar conditions of strains and temperature. This can be attributed to the fabrication technique which using PEDOT: PSS aqueous solution as a coagulant, where the phase inversion process is carried out, and in turn, thin layer of PEDOT: PSS is covering the nanofibers and sometimes in between the nanofibers.

(a) (b)

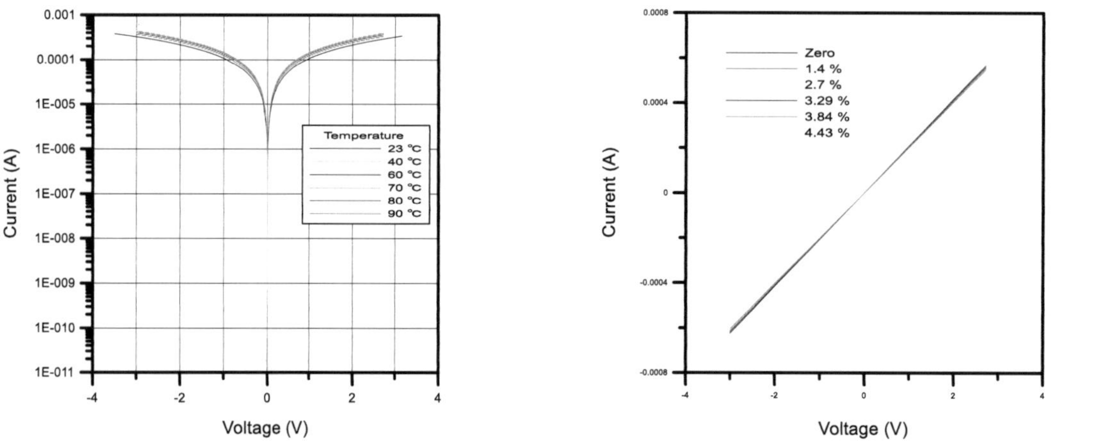

Figure 5. I-V characterization curves the as-spun TPU/PEDOT: PSS nanofiber membrane under (a) different temperatures (b) different strains.

4. Conclusions

In summary, the fabricated TPU/PEDOT: PSS nanofibrous membrane showed a great enhancement in electrical conductivity at KΩ scale rather than common MΩ scale in the previous studies. The mat demonstrates an ohmic behaviour irrespective the applied strain and the current decreases in a monotonic way with increasing the applied strain. On the other hand, current-voltage characteristics of the TPU/PEDOT: PSS nanofiber membrane under different percentages of tensile strain demonstrate an ohmic behaviour irrespective the applied strain and the current decreases in a monotonic way with increasing the applied strain. The experimental results are showing that the producd nanofibrous membrane is a promising candidate for strain sensing applications.

5. References

[1] Chen X, Zheng X, Kim J.-K, Li X and Lee D.-W 2011 *J. Vac. Sci. Technol.* B **29** 0601
[2] Chen Z, Ren W, Gao L, Liu B, Pei S and Cheng H M 2011 Nat. Mater 10 424

[3] Choong C L, Shim M B, Lee B S, Jeon S, Ko D S, Kang T H, Bae J, Lee S H, Byun K E, Im Jeong Y J, Park C E, Park J J and U. I. Chung 2014 *Adv. Mater* **26** 3451
[4] Lee J, Kim S, Lee J, Yang D, Park B C, Ryu S and Park I 2014 *Nanoscale* **6** 11932
[5] Li X, Zhang R, Yu W, Wang K, Wei J, Wu D, Cao A, Li Z, Cheng Y, Zheng Q, Ruoff R S and Zhu H 2012 *Sci. Rep.* **2** 870
[6] Liang H-W, Guan Q-F, Zhu Z, Song L-T, Yao H-B, Lei X and Yu S-H 2012 *NPG Asia Mater.* **4** 19
[7] Mannsfeld S C, Tee B C, Stoltenberg R M, Chen C V, Barman S, Muir B V, Sokolov A N, Reese C and Bao Z 2010 *Nat. Mater* **9** 859
[8] Park M. A. A. P. S. L. S. R. I. 2014 *Acs Nano* **8** 5154
[9] Park M, Im J, Shin M, Min Y, Park J, Cho H, Park S, Shim M B, Jeon S, Chung D Y, Bae J, Park J, Jeong U and Kim K 2012 *Nat. Nanotechnol* **7** 803
[10] Hu D, Wang Q, Yu J, Hao W, Lu H, Zhang G, Wang X and Qiu L 2016 Highly Stretchable Strain Sensors Using an Electrospun Polyurethane Nanofiber/Graphene Composite *Journal of Nanoscience and Nanotechnology* **16** 6 pp 5839-5842
[11] Sayed S, Gamil M, Fath El-Bab A M R and A. El-Moneim A 2015 Laser Reduced Graphene Oxide Film on Flexible Substrate for Strain Sensing Applications: Temperature Effect on Gauge Factor *Key Engineering Materials* **644** pp 115-119
[12] Gamil M, Nageh H, Bakrey I, Sayed S, Tabata O, Nakamura K, Fath El-Bab A M R and A. El Moneim A 2014 New Flexible Graphene-based Strain Gauge Sensor *Sensors and Materials* **26** 9 pp 699–709
[13] Sayed S, Fath El-Bab A M R, Nakamura K and A. El-Moneim A 2016 Development of Graphene Film on Flexible Poly-ethylene Terephthalate Substrate Using High Power Laser: Temperature Dependency of Gauge Factor for Graphene-based Strain Sensor *Sensors Review* **36** 2 pp 140-147
[14] Gamil M, Nakamura K, Fath El-Bab A M R, Tabata O and A. El-Moneim A 2014 Investigation of a New High Sensitive Micro-Electromechanical Strain Gauge Sensor Based on Graphene Piezoresistivity *Key Engineering Materials* **605** pp207-210
[15] Long Y Z, Yu M, Sun B, Gu CZ and Fan Z Y 2012 Recent advances in large-scale assembly of semiconducting inorganic nanowires and nanofibers for electronics, sensors and photovoltaics *Chem. Soc. Rev.* **41** 12 pp 4560–4580
[16] Schiffman J D and Schauer C L 2008 A review: electrospinning of biopolymer nanofibers and their applications *Polymer Rev.* **48** 2 pp 317–352
[17] Agarwal S, Wendorff J H and Greiner A 2008 Use of electrospinning technique for biomedical applications *Polymer* **49** 26 pp 5603–5621
[18] Ali A A and Al-Asmari A Kh 2012 Wet-electrospun CuNP/carbon nanofibril composites: Potential application for micro surface-mounted components *Applied Nano - science* **2** pp 55–61
[19] Luká D, Sarkar A, Martinová L, Vodsed'álková K, Lubasová D, Chaloupek J, Pokorn P, Mike P, Chvojka J and Komárek M. 2009 Physical principles of electrospinning (electrospinning as a nano-scale technology of the twenty-first century) *Textile Progress* **41** pp 59–140
[20] Zhong S, Zhang Y and Lim C T 2012 Fabrication of large pores in electrospun nanofibrous scaffolds for cellular infiltration: A review *Tissue Engineering Part B: Reviews* **18** pp 77–87
[21] Kostakova E, Seps M, Pokorny P and Lukas D 2014 Study of polycaprolactone wet electrospinning process. *Express Polymer Letters* **8** 8
[22] Seyedin M Z, Razal J M, Innis P C and Wallace G G 2014 Strain - Responsive Polyurethane/PEDOT: PSS Elastomeric Composite Fibers with High Electrical Conductivity
[23] Park M, Im J, Shin M, Min Y, Park J, Cho H, Park S, Shim M-B, Jeon S, Chung D-Y, Bae J, Jeong U and Kim K 2012 *Nat. Nanotechnol* **7** 803
[24] Liu N, Fang G, Wan J, Zhou H, Long H and Zhao X 2011 Electrospun PEDOT: PSS–PVA nanofiber based ultrahigh-strain sensors with controllable electrical conductivity *Journal of Materials Chemistry* **21** 47 pp18962-18966

[25] Polat Y, Pampal E S, Stojanovska E, Simsek R, Hassanin A, Kilic A, Demir A and Yilmaz S 2016 Solution blowing of thermoplastic polyurethane nanofibers: A facile method to produce flexible porous materials *Journal of Applied Polymer Science* **133** 9
[26] Jia Y T, Gong J, Gu X H, Kim H Y, Dong J and Shen X Y 2007 *Carbohydr Polym* **67** 403

Acknowledgements
This work was supported by Central Research Laboratories and Technology Management Department at Egypt-Japan University of Science and Technology, Alexandria, Egypt. The work was fully funded by Ministry of Higher Education (MOHE) of Egypt. Thanks are also for Eng. Enas Mostafa for her kind experimental and analytical effort in this work.

IOP Conf. Series: Materials Science and Engineering **244** (2017) 012011 doi:10.1088/1757-899X/244/1/012011

Preliminary investigation of fabrication composite structures by using shape memory alloys

W Klein[1] and O Dudek[1]

1 Faculty of Mechanical Engineering, Silesian University of Technology, PL
E-mail: wojciech.klein@polsl.pl

Abstract. The paper shows method of smart forming composite structures and fundamentals of propose fabrication technology. The presented method is based on innovative 3D printing technics with SMA (Shape Memory Alloy) fibres application. The SMA fibres layout cause an eccentric axial load after thermal activation. The result of this process is composite structures deflection in a predictable direction. The technology demonstrator sample was fabricated as well as numerical simulations were performed in aim of proof of concept. The identification process was developed to determine the layout of SMA fibres. The simulations were performed in MATLAB and ANSYS environment, where the genetic algorithm was used to identify geometrical parameters. The MAC (Modal Assurance Criterion) criterion was used to compare nodal solution with the predefined shape pattern. The simulation results shows possibilities of forming composite structures on the example of deflected beam.

1. Introduction

Shape memory alloys are a group of material, which are characterized by the ability to undergo a large deformation about 8% to 10% without permanent deformation effect [1-2]. If recovering to initial shape is self-contained that effect is called superelasticity or pseudoelasticity. If the above mentioned recovery is a result of heating to the specified temperature, the shape memory effect is observed then. These alloys are classified as smart materials and are frequently used to fabricate smart composite structures [3-8].

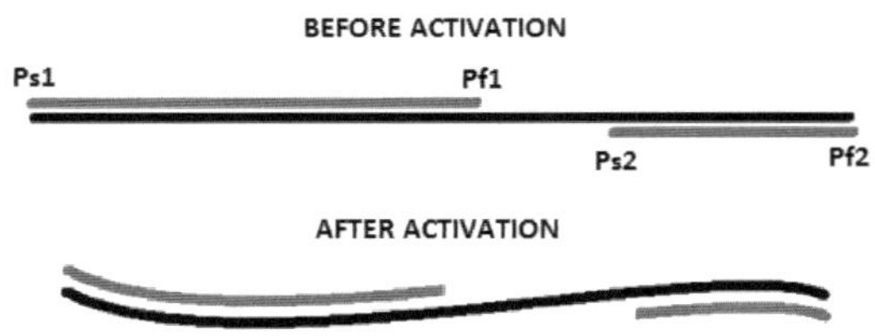

Figure 1. Example of SMA fibre fixing and expected beam shape after activation.

Proposed method of smart fabrication of composite structures is based on thermal actuation of SMA fibers. After activation SMA fibers are contracted and induce force which deforms structure in appropriate manner (Figure 1). Obtaining a predefined shape depends on many factors. The main assumption of this method is reducing costs and accelerating the manufacturing process of composite structures. In this aim SMA fibers layout was applied by using 3D printing method. The Figures 2 and 3 present example of 3D printed composite structure before and after activation.

IOP Conf. Series: Materials Science and Engineering **244** (2017) 012011 doi:10.1088/1757-899X/244/1/012011

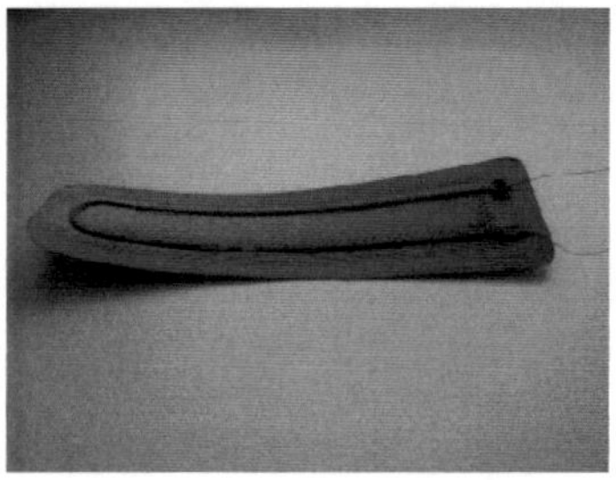

Figure 2. Sample actuator made by 3D printing before activation.

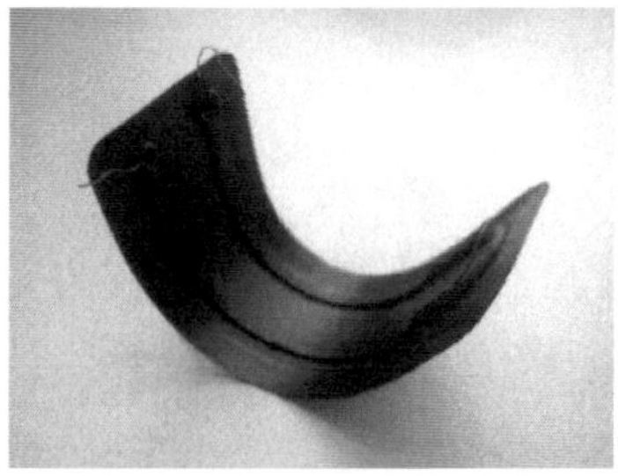

Figure 3. Sample actuator made by 3D printing after activation.

2. Theoretical background

A simplified problem of composite structures deflection is a blending of simple beam. Description of the simple beam bending phenomena can be find for example in literature [9]. The main assumption which allow to analyze this task is an eccentric axial load that not act through the centroid of the beam cross section. An example is shown in Figure 4, where the cantilever beam Pp – Pk is subjected to a tensile load P acting at distance D from the X axis.

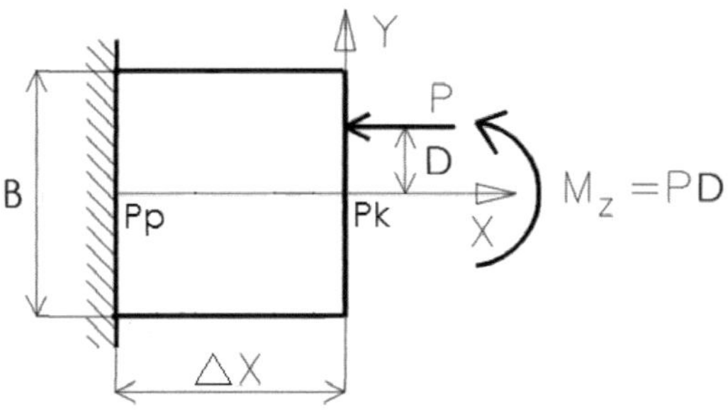

Figure 4. Simplified model of an eccentric load beam.

The eccentric force P is statically equivalent to an axial force P acting along the x axis and a bending moment M_b acting about the perpendicular axis.

$$M_b = P \times D \tag{1}$$

where:

M_b – bending moment
P – eccentric tensile force
D – distance from the neutral axis

According to method of superposition the slope or deflection at any point on the beam is equal to the resultant of the slopes or deflections at that point caused by each of the load acting separately. The maximum deflection of the cantilever beam is equal:

$$\delta_{Pf} = \frac{M_b \cdot \Delta x^2}{2E \cdot I_Z} \tag{2}$$

where:

δ_{Pf} – beam deflection in the Pf point
Δx – distance from point Ps to point Pf
E – Young's modulus
I_Z – moment of inertia of plane area

3. Numerical simulation

The structure model has been simplified to a one-dimensional beam with a fixed defined cross-section. The beam was divided into 100 finite elements of the size of 1 mm each. The finite element type used in the analysis was BEAM188 and is suitable for analyzing slender and medium thick beam structures. This element is based on Timoshenko's beam theory, which includes shear effects.

One of the main simplifications in numerical simulations was the replacement of the SMA material model with a model which assuming adequate contraction to the SMA fiber according to the temperature. This dependency was made by applying negative thermal expansion coefficient of SMA fiber model.

The major aim of numerical simulation was identification of geometrical parameters of composite structure. These parameters are related with position of SMA fibers and their length. The developed identification process was based on genetic algorithms. The fitness function compared simulation results of particular iteration. The MAC (Modal Assurance Criterion) was used to relate degree of consistency between reference pattern shape and actual nodal solution [10].

$$MAC\ (sp, ns) = \frac{|\{\varphi_{sp}\}^T\{\varphi_{ns}\}|^2}{\left(\{\varphi_{sp}\}^T\{\varphi_{sp}\}\right)\left(\{\varphi_{ns}\}^T\{\varphi_{ns}\}\right)}, \quad (3)$$

where:

MAC – Modal Assurance Criterion
φ_{sp} – vector of shape pattern nodes
φ_{ns} – vector of actual nodal solution

The ANSYS solver was used to calculate parametric finite element model and transfer results in matrix form to a specify file. The MATLAB solver read results from the file and run genetic algorithm optimization process to identify composite structures geometrical parameters. The flowchart of this process is shown in Figure 5.

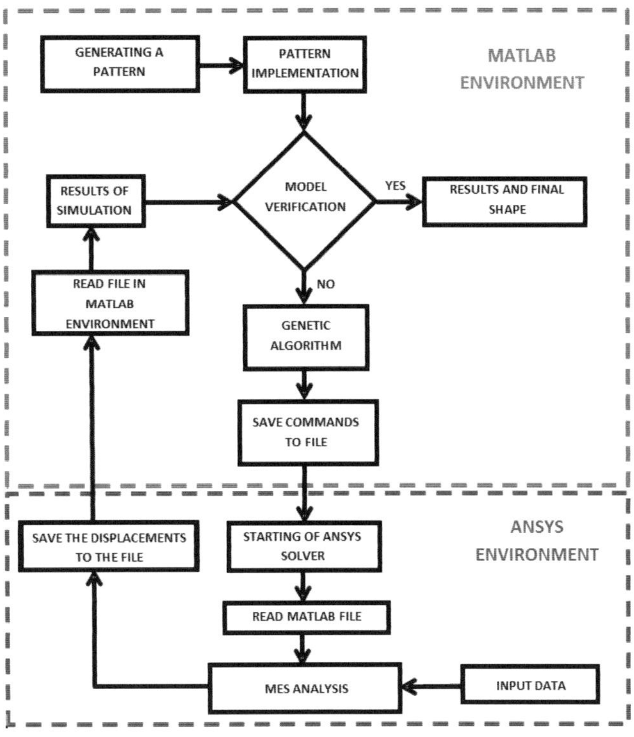

Figure 5. Flowchart of the identification process.

4. Result of simulations

The simulations were performed for three different pattern shapes. For each of them are presented fitness function values and shape changes dependent on iteration numbers.

The simulation for "sinus" pattern shape

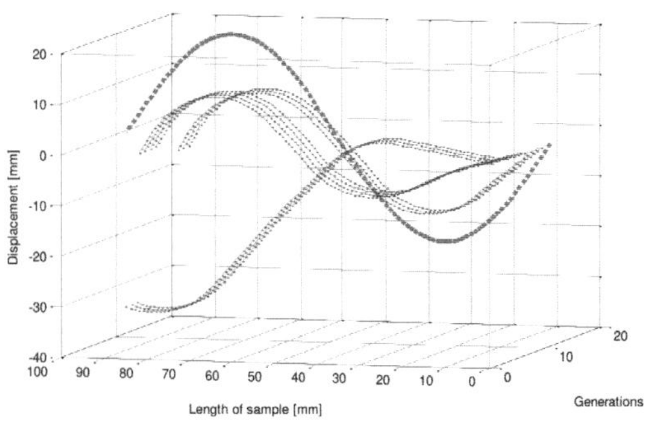

Figure 6. Shape changes depend on the iterations numbers.

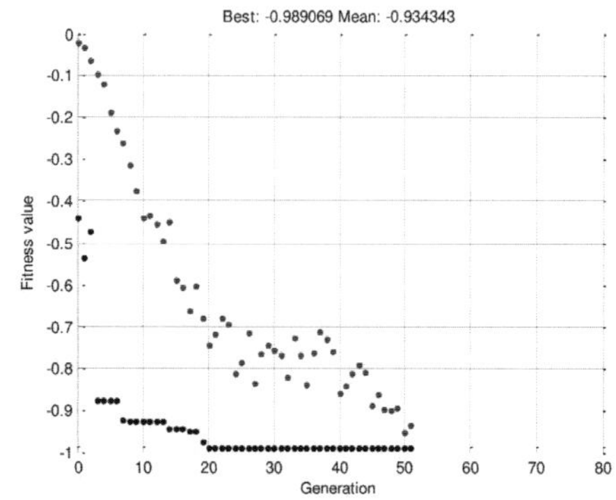

Figure 7. Average and maximum values of the fitness function for each generation.

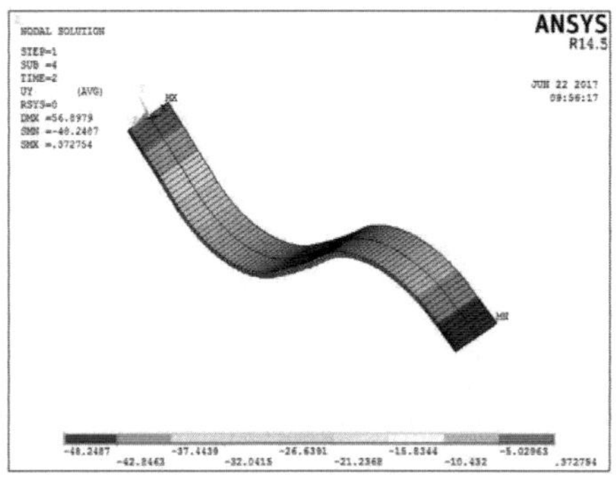

Figure 8. Node's displacements of final iteration.

The simulation for "U" pattern shape

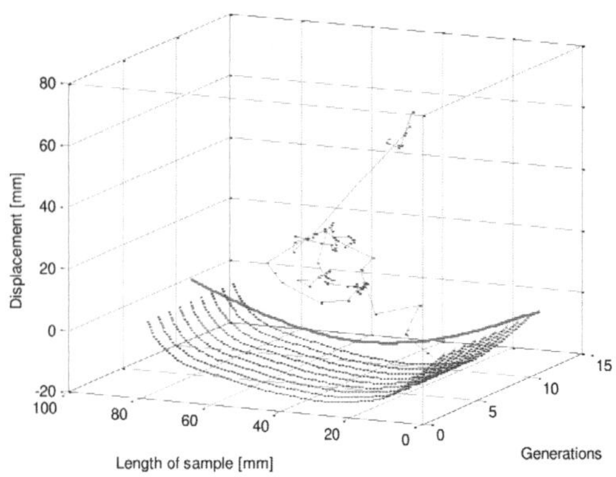

Figure 9. Shape changes depend on the iterations numbers.

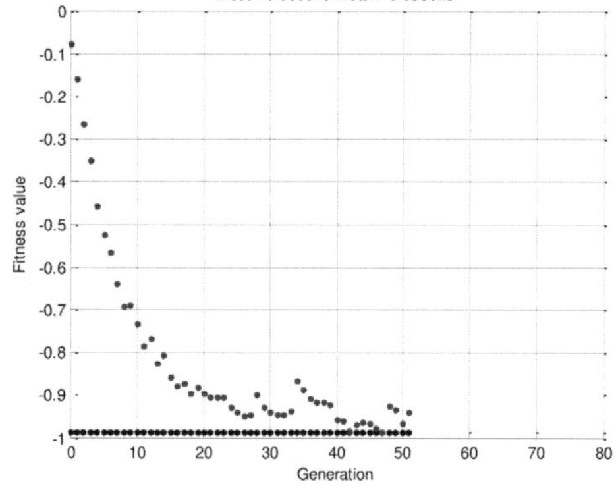

Figure 10. Average and maximum values of the fitness function for each generation.

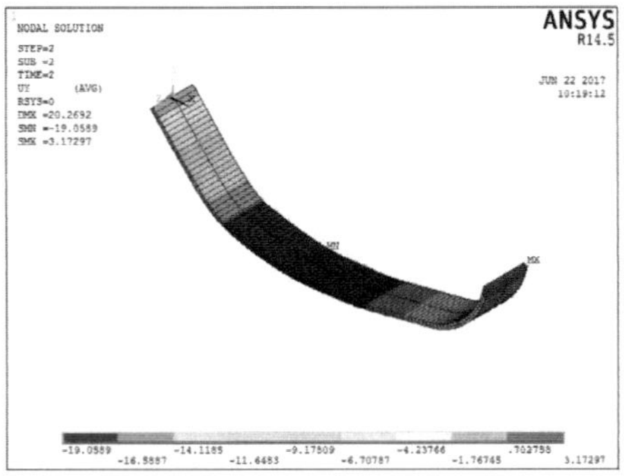

Figure 11. Node's displacements of final iteration.

The simulation for "trapezium" pattern shape

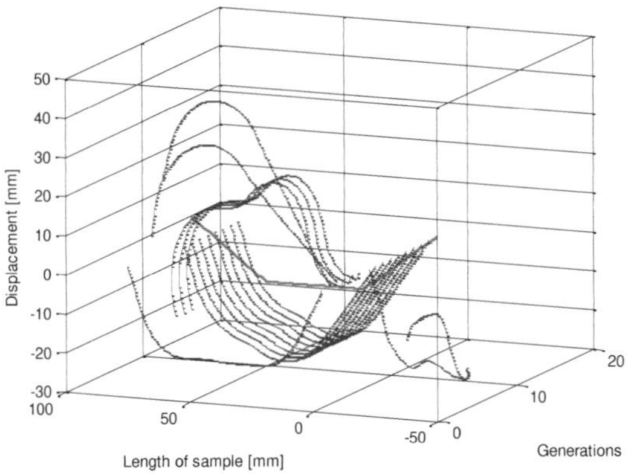

Figure 12. Shape changes depend on the iterations numbers.

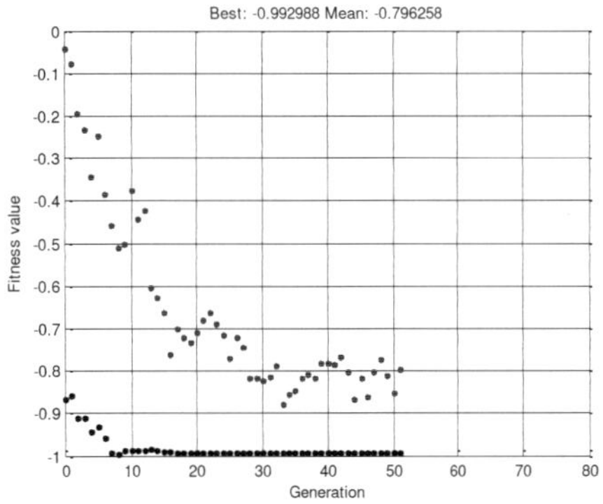

Figure 13. Average and maximum values of the fitness function for each generation.

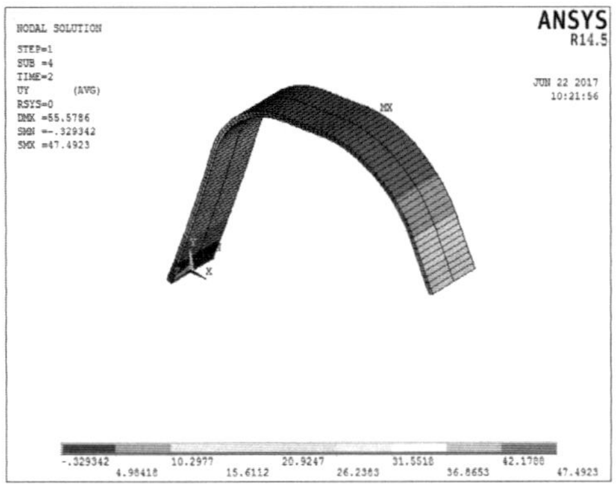

Figure 14. Node's displacements of final iteration.

The summary of simulation results is presented in Table 1. All simulations didn't exceeds 52 generation of genetic algorithm and had similar fitness function values.

Table 1. The summary of simulation results.

	Fitness Value	Number of generation	Point *Ps1*	Point *Pf1*	Point *Ps2*	Point *Ps2*
Shape pattern "sinus"	0,989	52	17	44	63	93
Shape pattern "U"	0,988	52	78	100	28	54
Shape pattern "trapezium"	0,993	52	33	56	66	90

5. Conclusions

The presented in this paper method of smart fabrication of composite structures requires further research works. There are several main aspects which could improve the proposed technology. One of them is obtaining more accurate shape from shape pattern by using more SMA fibers. This is related to increasing the number of variables in the identification process, which will affect the timing of the individual simulations. It is also important to check the possibility of forming spatial shapes using appropriately developed discrete models.

Another possibility for the development of research is to perform experimental research to verify the numerical models and to improve the production technology of such structures. It will be a great challenge to accurately positioning of SMA fibers relative to the neutral axis in the bending beam or shell. This parameter has significantly influence on degree of beam deflection. The solution to this problem can be a special 3D printer construction with double head – first for polymer matrix and second for SMA fiber application.

6. References

[1] Otsuka K and Wayman C, 1999 Mechanism of shape memory effect and superelasticity *Shape memory materials*, 1st ed., Cambridge: Cambridge University Press, pp 27–48

[2] Bellouard Y 2002 Microrobotics, microdevices based on shape memory alloys *Encyclopedia of smart materials*, New York: John Wiley and Sons Inc. pp 620-643

[3] Wei Z, Sandstrorӧm R, and Miyazaki 1998 Shape-memory materials and hybrid composites for smart systems: Part I Shape-memory materials *Journal of Materials Science* Volume 33, Issue 15, pp 3743–3762

[4] Lester B, Baxevanis T, Chemisky Y and Lagouda D 2015 Review and perspectives: shape memory alloy composite systems, *Acta Mechanica* Volume 226, Issue 12, pp 3907–3960

[5] Aboudi J 1997 The response of shape memory alloy composites *Smart Materials and Structures* Volume 6, Number 1

[6] Ni D and Ma Z 2014 Shape Memory Alloy-Reinforced Metal-Matrix Composites: A Review *Acta Metallurgica Sinica* Volume 27, Issue 5, pp 739–761

[7] Asadi H, Bodaghi, M, Shakeri, M and Aghdam M 2014 Nonlinear dynamics of SMA-fiber reinforced composite beams subjected to a primary/secondary-resonance excitation *Acta Mechanica* Volume 226, Issue 2, pp 437–455

[8] Mężyk A, Klein W, Pawlak M and Dudek O 2016 Preliminary tests of new composite material based on shape memory alloys for artificial aortic valve leaflets *International Conference on Bio-engineering for Smart Technologies. BioSMART 2016*, Dubai, (UAE)

[9] Gere J and Goodno B 2013 *Mechanics of Materials Cengage Learning*

[10] Allemang R 2003 The Modal Assurance Criterion –Twenty Years of Use and Abuse *Sound & Vibration* pp 14-21

Acknowledgments

The publication has been financed from the statutory grant of the Faculty of Mechanical Engineering of the Silesian University of Technology in 2017.

IOP Conf. Series: Materials Science and Engineering **244** (2017) 012012 doi:10.1088/1757-899X/244/1/012012

Wear properties of 10 vol.% silicon carbide particulate-reinforced aluminum composite fabricated by powder injection molding

T Patcharawit[1], A Ngeekoh[2] and N Chuankrekkul[3]

1 Lecturer, School of Metallurgical Engineering, Suranaree University of Technology, Nakhon Ratchasima, Thailand
2 Master student, School of Metallurgical Engineering, Suranaree University of Technology, Nakhon Ratchasima, Thailand
3 Lecturer, The Metallurgy and Materials Science Research Institute, Chulalongkorn University, Bangkok, Thailand
E-mail: tapany@sut.ac.th

Abstract. Wear properties of aluminum matrix composites reinforced with silicon carbide particulate of 10 vol.% addition was investigated in as-sintered and heat-treated conditions under varying loads at -5, -25,- 45 and -65N using a ball on flat type of wear test. The composite was fabricated by powder injection molding and sintering at 650 °C for 3 hours. Solution treatment was carried out at 550 °C for 2 hours followed by age-hardening at 160 °C for 6 hours. SEM and XRD results indicated Al and SiC_p are present as matrix and reinforcement, while AlN, Al_2Cu and Mg_2Si were also detected. Further precipitation of Al_2Cu and Mg_2Si in heat-treated samples promoted maximum macro and micro Vickers hardness values, which were achieved at 161 and 157 H_v respectively. Wear weight loss increased with increasing minus load level. The coefficient of friction was found in the range of 0.042-0.048. Wear mechanisms were determined as the combination of abrasive, adhesion and oxidation.

1. Introduction

Aluminum matrix composites have been increasingly used for engineering applications such as automotive, aerospace, biomedical and electrical applications [1]. Light weight, good hardness and strength as well as wear resistance are attractive properties, which are controlled by starting materials used and fabrication techniques. Silicon carbide particulate as the reinforcing material provides isotropic properties, if good distribution of the reinforcement and effective bonding between the reinforcement and the matrix are attained [2]. Strengthening mechanisms were applied to improve mechanical properties of aluminum composite such as solid solution and precipitation hardening.
In recent years, it is known that small addition of silicon carbide reinforcement can improve hardness of the aluminum matrix. However, the efficiency of reinforcements are not good enough for hardness and tensile strength due to the different hardness between reinforcement and matrix [2]. Further, effective interfacial bonding between aluminum matrix and silicon carbide strongly determines the final properties, which depends on selected processing routes. Research on stir-mixing, liquid infiltration and powder technology of aluminum composites are under intense investigation to enhance properties and scope engineering applications [3]. However, there are certain limitations of liquid-state forming processes such as wettability of aluminum and silicon carbide, and brittle aluminum carbide (Al_4C_3) occuring along interfaces between aluminum matrix and silicon carbide. Powder metallurgy

however has advantages such as low-cost for mass production of small parts, uniform distribution of particle reinforcements, and solid state process to reduce the reactions between the matrix and particles [4]. Though, studies on conventional powder metallurgy of aluminuim composites have been increasing published, powder injection molding (PIM) of aluminum composites on the other hand are quite limited. With versatility of the PIM in producing small complex shaped components at low cost, it is also a good candidate for the production of aluminum composite with uniform reinforcement distribution.

Microstructure and properties such as hardness and tensile have been documented elsewhere [5]. In this research, wear properties of 10 vol.% silicon carbide reinforced aluminum composite has been studied to extend the possibility of using aluminum composite subjected to engineering wear. The research aim is to study the influences of different loads on wear behaviour of powder injection molded silicon carbide reinforced aluminum composite after sintering and precipitation hardening under ball on flat type of wear test.

2. Experimentals

Aluminum powder and silicon carbide particulate reinforcement were used as the feedstock. The compositions of aluminum alloy powder (by weight percent) is Cu 4.3-4.7, Mg 0.4-0.6, Si 0.5-0.8, Fe 0.07 and balanced Al. Silicon carbide particulate was added at 10 vol.% to the aluminum powder prepared by ball-mill mixing for 2 hours at 280 rpm. before mixing with a multi-component polymeric binder. The mixed powders of 55 % solid loading were injection molded at 170 °C and 45 MPa to produce green samples of 20x20x5 mm^3 in dimensions. After debinding in hexane, the brown samples were sintered at 650 °C in a high purity nitrogen atmosphere. Solution heat treatment was carried out at 550 °C for 2 hours. followed by age-hardening at 160 °C for 6 hours. Microstructure (OM and SEM), bulk density and wear resistance were examined for the as sintered and heat-treated samples. Ball on flat type of wear test according to ASTM G 133-95 were employed, using load varying at -5, -25, -45, and -65N, for a sliding distance of 100 m. with no lubrication. The 440-C stainless steel balls of 6.3 mm. diameter, having 62 HRC hardness were used as the counterface material.

3. Results and discussion

3.1. Microstructure and phase analysis

Sintered microstructure of the aluminum composite exhibits uniform distribution of silicon carbide particulate in the aluminum matrix in the as-sintered condition, as shown in Figure 1 a). Some areas contain micro-porosity along silicon carbide clusters. SEM micrograph at 1000X magnification reveals precipitate phase of Al_2Cu after heat treatment, as shown in Figure 1 b), confirmed by EDS analysis.

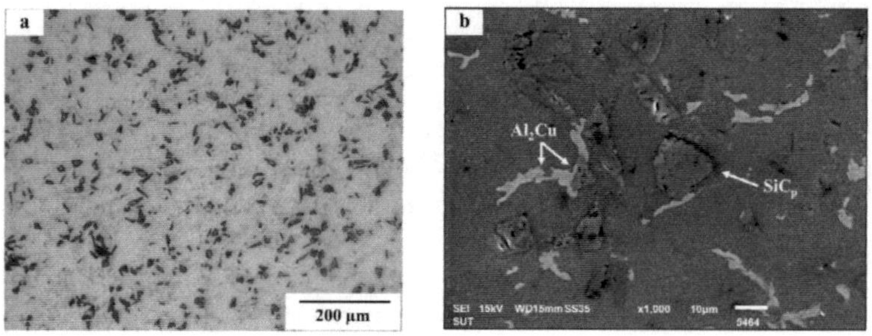

Figure 1. Microstructures showing a) silicon carbide distributed in aluminum matrix (OM) and b) at higher magnification (SEM)

EDS mapping analysis, as shown in Figure 2, indicates the aluminum matrix contains main elements such as Al and Cu. Si was found where silicon carbide particulate is located. XRD results of the aluminum composite after sintering and heat treatment are demonstrated in Figures 3 a) and b) respectively. The main phases are aluminum matrix and silicon carbide. It is noted that AlN was detected in both of sintered and heat-treated conditions due to the effect of nitrogen gas in sintering atmosphere. Furthermore, there are higher peaks of precipitate phases such as Al_2Cu and Mg_2Si found in the heat-treated composites, which contribute to improved hardness.

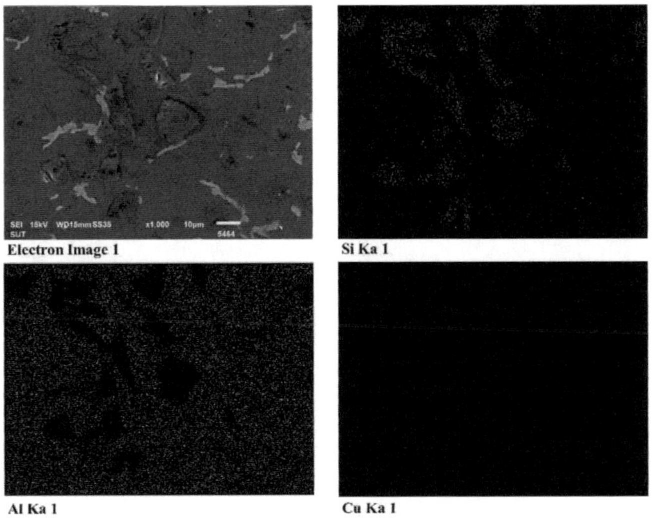

Figure 2. EDS mapping analysis of Al, Si and Cu.

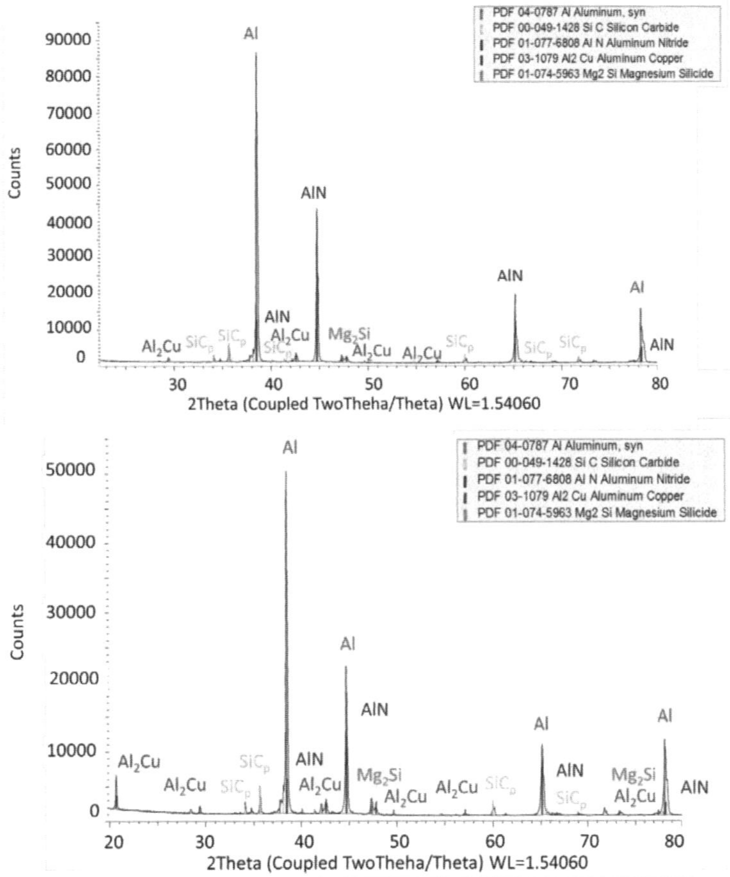

Figure 3. XRD result of silicon carbide particulate reinforced aluminum composite after sintering (top) and precipitation hardening (bottom).

3.2. Density and hardness

Density was measured to be 1.82 g/cm^3 after injection molding and then increased to 2.63 g/cm^3 after sintering. This accounts for 93.2% theoretical density in the sintered condition according to the rule of mixture by taking the densities of aluminum alloy = 2.78 g/cm^3 and silicon carbide = 3.20 g/cm^3 respectively. The sintered density, which is lower than the theoretical density is plausibly due to porosity left after sintering, as previously shown in Figure 1 a).

Macro and micro Vickers hardness values of heat-treated samples significantly increase as compared to those of as-sintered samples, as shown in Table 1. The maximum values were measured at 160.1 and 156.9 Hv respectively. This is due to precipitation hardening of Al_2Cu and Mg_2Si in the matrix.

Table 1. Macro and micro Vickers hardness values for as-sintered and heat-treated conditions.

	Macro Hardness (Hv)	Micro Hardness (Hv)
As sintered	130.1	136.0
Heat treated	160.1	156.9

3.3. Wear properties and worn surfaces

Wear grooves at varying minus loads at a stroke length of 10 mm are depicted in Figure 4. Increasing wear groove width is observed with increasing minus load level. Furthermore, plastic deformation at the edges of the wear grooves along the wear track at higher minus load level is evident.

Wear test parameters such as frictional force (Fx) and normal force (Fz) were measured in real time as shown in Figures 5 a) and b) for loads at -25 and -65N respectively, presented at t = 0 – 400 second. The coefficient of friction (COF) was then obtained accordingly.

Wear weight loss as a function of load for as-sintered and heat-treated samples are illustrated in Figures 6 a) and b) respectively. It is noted that wear weight loss at -5N load for both conditions is undetectable. It can be seen that a slight increase trend of wear weight loss was observed with increasing minus load [6]. The COF are reported in Figures 7 a) and b), and was found in a range of 0.042 - 0.048 and 0.044 - 0.047 for as-sintered and heat-treated conditions respectively. These values are comparable to those observed in laser composite surfaced aluminum with silicon carbide by J. Dutta Majumdar et. al, [7]. Many researches [8] also supported that the COF values were found insensitive to load or pressure applied. However, COF at very low minus load of -5N was significantly higher. Further study is required for more explanation.

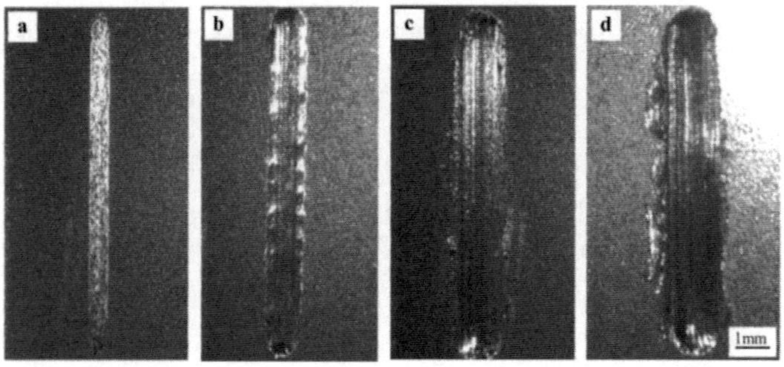

Figure 4. Wear grooves at varying loads at a) -5N, b) -25N, c) -45N and d) -65N.

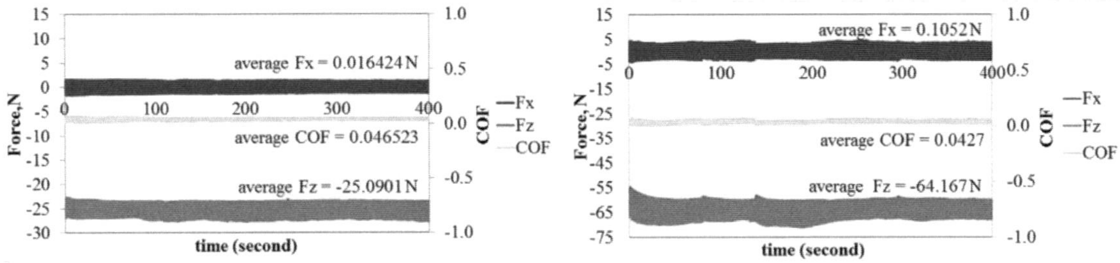

Figure 5. Test parameters at a) -25N and b) -65N, showing Fx, Fz and coefficient of friction (COF).

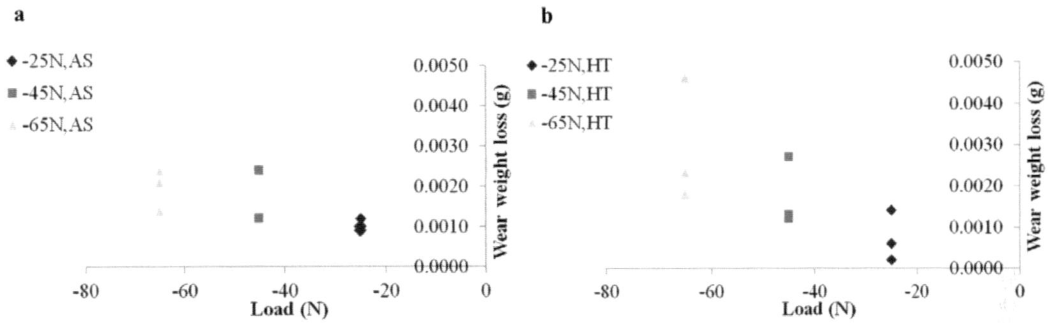

Figure 6. Wear weight loss as a function of load in a) as-sintered and b) heat-treated conditions.

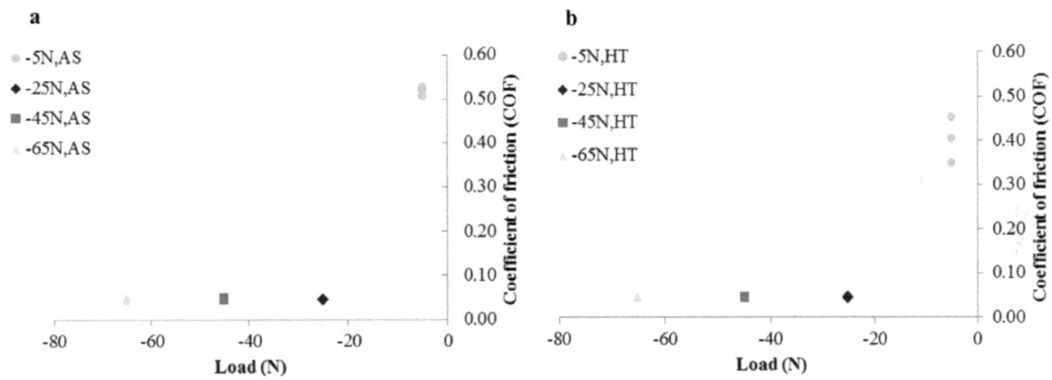

Figure 7. Coefficient of friction as a function of load in a) as-sintered and b) heat-treated conditions.

Worn surface of aluminum composite were investigated by using SEM, as demonstrated in Figures 8 a) and c) at low magnification for -25N and -65N respectively. Abraded surfaces are evident in all samples at every load levels, indicating abrasive wear, but with a lower degree at -5N. As the counterface stainless steel ball is much harder in comparison to the aluminum composite, it is believed that silicon carbide might have been pulled out and abraded on the softer aluminum matrix. The increasing contact load between stainless steel ball and the aluminum composite had made the latter to deform progressively, as shown in Figure 4. Secondary electron micrographs at higher magnification are shown in Figures 8 b) and d). Rough surface area of wear grooves are possibly due to deformation of the soft aluminum matrix. Severely deformed matrix would then be delaminated, transferred and adhered onto other surface. Back-Scattered Electron mode analysis as depicted in Figure 9 a) indicates certain amount of silicon carbide distributed throughout the aluminum matrix. However, BSE micrograph of the worn surface at the same magnification reveals only small traces of silicon carbide left on the surface, as seen in Figure 9 b). It is then confirmed that some silicon carbide particulate previously embedded in the aluminum matrix fell out and might have abraded on to the softer aluminum matrix during the ball on flat wear test, introducing three body abrasion. Similar results by F. Alshmri, et. al. [9] and I. Dinaharan, et. al. [10] indicated harder particles (silicon or fly ash) being pulled out and promoted three body abrasion wear, which then created fine spherical wear debris [9, 10]. Though, silicon carbide fracturing was also possible in this research at higher load, the evidence of fractured silicon carbide was however hardly observed. Therefore, it is worth to investigate effects of silicon carbide content on the degree of abrasive wear on this composite in future work.

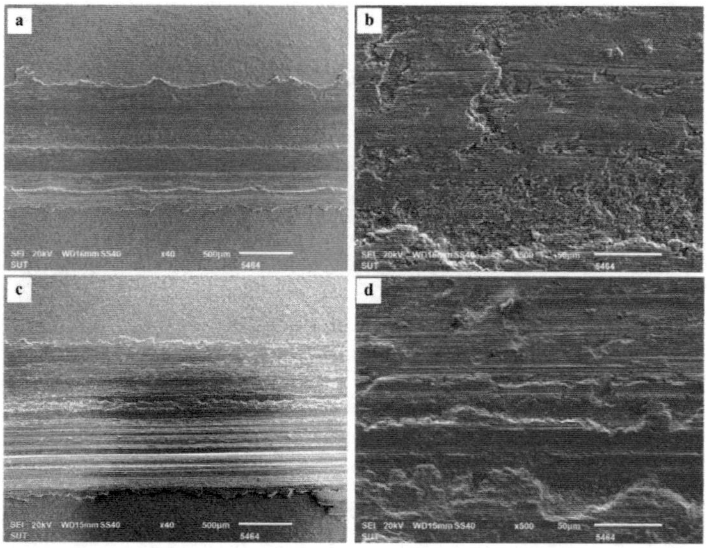

Figure 8. Secondary electron micrographs of worn surfaces at a) -25N and b) higher magnification, c) -65N and d) higher magnification.

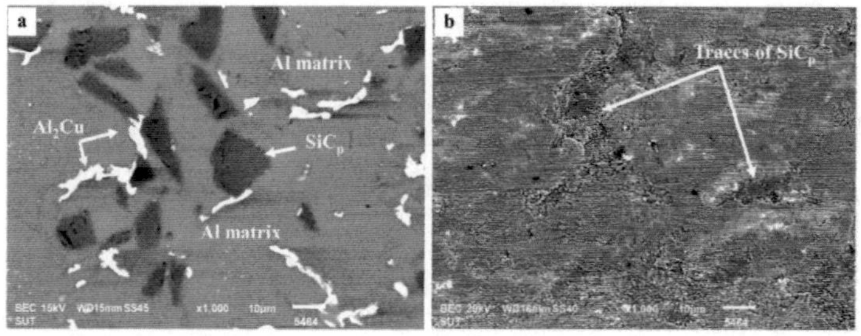

Figure 9. Back-scattered electron micrograph a) before wear test and b) after wear test.

Wear debris obtained from all tests at varying minus loads have shown two different forms of agglomerate and flake like-shape. EDS point analyses of wear debris at load levels of -25N and -65N, are illustrated in Figures 10 a) and b) respectively. Peaks of Al, Mg, Cu and O elements were detected in spectrum 1 of agglomerate wear debris. Flake like-shaped wear debris was EDS analysed (spectrum 2), showing Al, Mg, Cu and O elements having different composition in comparison to that of agglomerate one (spectrum 1). EDS analysis on wear debris obtained from lower minus load test indicates higher content of Al on the flake-like form of wear debris, which signifies that this product might be from the deformed aluminum matrix. Moreover, higher level of oxygen observed in the agglomerate form suggests more of oxidation product. At higher level of minus load (-65N), the level of Al and O content detected in both agglomerate and flake-like forms are comparable, as shown in Figures 10 a) and b). It is therefore believed that oxidation wear might have occurred in a mild level during the wear test and is more significant at higher minus load level [11]. Moreover, there was however no peaks of Fe, Ni or Cr detected. This implies the mutual transfer of material between the

wearing aluminum composite and the stainless steel counterface appears to be insignificant. Only aluminum matrix was deformed, oxidized, detached and transferred to be adhered on other areas.
In addition, it is noticed that the size of flake-like form of wear debris also became significantly larger with increasing minus load level. Dry sliding wear test on aluminum-high silicon hypereutectic alloys by F. Alshmri, et.al [9] indicated comparable result, where the transition from powder-type to flake like-shape as the load increased was reported. This flake-like wear debris found in this research also confirmed the operating wear mechanism is related to adhesive [10]. Together with the appearance of plastic deformation at the wear groove edges along the wear track previously mentioned, analogous to those observed from works by B. Hekner, et. al. [12] and S. Sawla and S. Das [13], these evidences substantiates the influence of plastic deformation on wear mechanisms at higher minus load level in this ball on flat type of wear test.

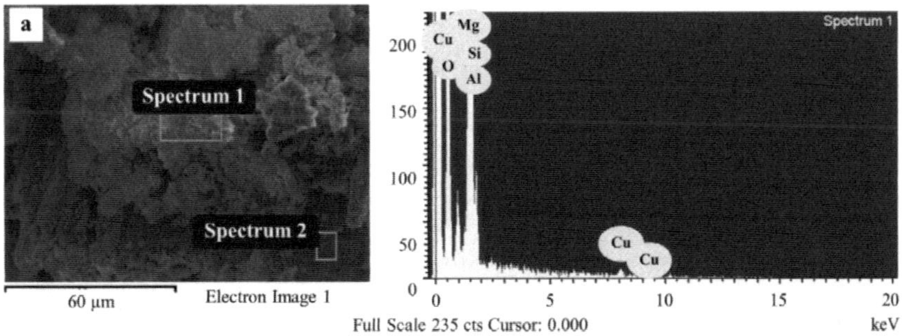

Element	Spectrum 1	Spectrum 2
O	46.02	29.77
Mg	0.64	0.82
Al	44.59	61.82
Si	3.12	1.60
Cu	5.63	5.99

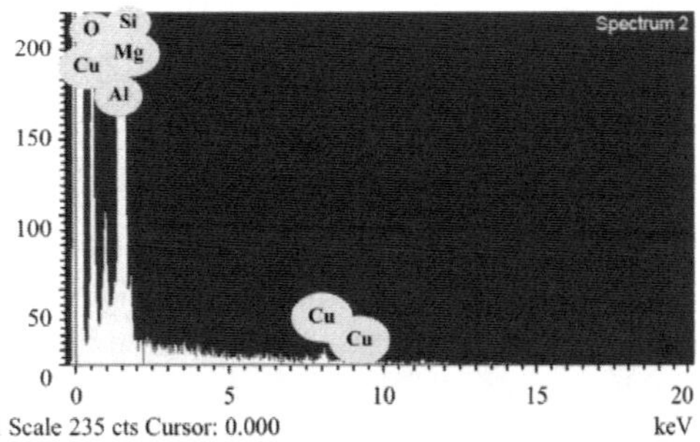

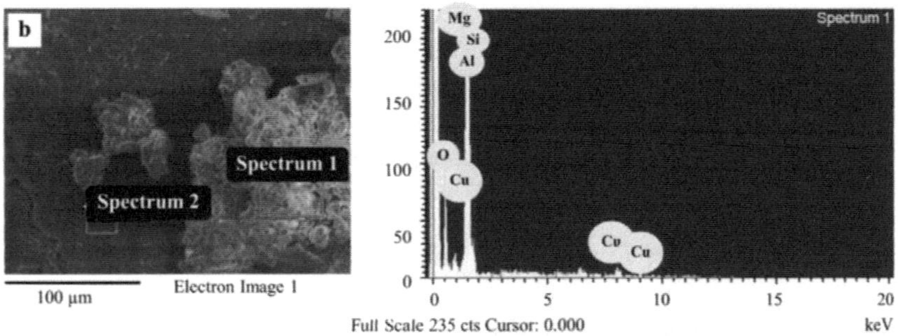

Element	Spectrum 1	Spectrum 2
O	45.95	40.96
Mg	0.89	0.67
Al	47.45	50.18
Si	2.62	2.66
Cu	3.09	5.53

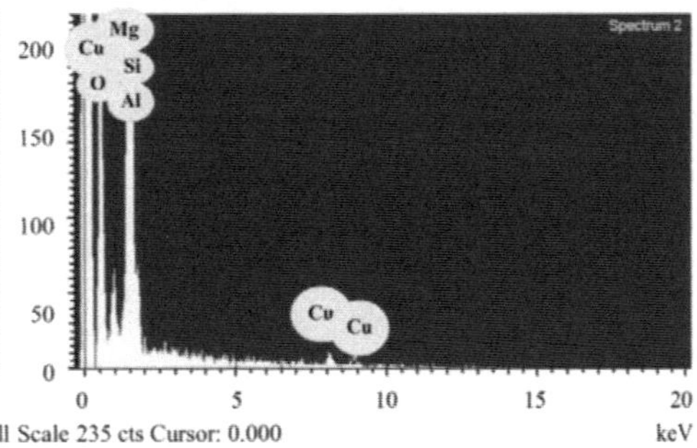

Figure 10. EDS point analysis of wear debris at aplied loads of a) -25N and b) -65N.

Therefore, wear mechanisms of aluminum matrix reinforced with 10 vol.% silicon carbide particulate fabricated by powder injection molding tested under ball on flat type of wear test with load varying at -5 to -65N can be determined as follows. Wear damage of the aluminum composites shows several wear mechanisms such as abrasive, adhesive and oxidation wear. Firstly, abrasive wear of aluminum reinforced with silicon carbide is caused by different hardness between the aluminum matrix and silicon carbide. Much harder silicon carbide particulate first fell out from the matrix and abraded onto the softer matrix surface. Secondly, aluminum matrix was deformed and delaminated to give wear debris and some was then transferred from one surface to another. Adhering of these transferred materials at the wear groove edge is also evident. Increasing minus load level introduced much more severity of the wear damage. Finally, it is known that aluminum passively produces oxide film layer on its fresh surface. During ball on flat of wear test in ambient atmosphere, the oxide film layer was removed and the fresh surface will react with oxygen immediately to form new oxide layers in repeated cycles of reciprocating wear test.

4. Conclusions

Wear mechanisms of powder injection molded aluminum composite reinforoced with 10 vol.% silicon carbide is due to abrasive, adhesive and oxidation. Abrading of silicon carbide on the softer aluminum matrix is due to hardness difference between the two materials. Mutual transfer between the two mating materials is not apparent. Only softer alumium matrix was plastically deformed, oxidized, detached and transferred to be adhered on other areas. Severe wear damage at high applied minus load of -65N involved plastic deformation of the soft matrix at the wear groove edges. Degree of damage such as wear weight loss, wear groove width and oxidation increased with increasing minus load level. The coefficient of friction has not significantly changed with the load applied and was found in the range of 0.042-0.048. However, wear damage of the composite after sintering and precipitation hardening are not significantly different.

5. References

[1] Sahin Y 2010 *Tribol Int* **43** 939–43
[2] Wang H, Zhang R, Hu X, Wang C and Huang Y 2008 *J. Mater. Process. Technol.* **197** 43–48
[3] Eslamian M, Rak J and Ashgriz N 2008 *J. Powder Technol.* **184** 11–20
[4] Martin M J and Castro F 2003 *J. Mater. Process. Technol.* **143–44** 814–21
[5] Patcharawit T, Ngeekoh A, Tongkerd W, Takhampom S, Lapkeaw S and Chuankrerkkul N 2017 *Key Eng. Mater.* **751** 60-69
[6] Nassar E A and Nassar E E 2017 *JKSUES* **29** 295-99
[7] Majumdar D J, Chandra R B and Manna I 2007 *Wear* **262** 641-48
[8] Deuis L R, Subramanian C and Yeilup M J 1997 *Compos. Sci. Technol.* **57** 415-35
[9] Alshmri F, Atkinson V H, Hainsworth V S , Haidon C and Lawes S D A 2014 *Wear* **313** 106-16
[10] Dinaharan I, Nelson R, Vijay J S and Akinlabi T E 2016 *Mater. Charact.* **118** 149-58
[11] Singh J and Chauhan A 2016 *Ceram. Int.* **42** 56-81
[12] Hekner B, Myalski J, Valle N, Probierz A B, Lizer S M and Wieczorek J 2017 *Composites Part B* **108** 291-00
[13] Sawla S and Das S 2004 *Wear* **257** 555-61

Acknowledgments

Authors gratefully acknowledge School of Metallurgical Engineering, Suranaree University of Technology, The Metallurgy and Materials Science Research Institute, Chulalongkorn University, and National Metal and Materials Technology Center for facilities and technical supports. Thanks are also due to Thailand Institute of Scientific and Technological Research for wear testing.

IOP Conf. Series: Materials Science and Engineering **244** (2017) 012013 doi:10.1088/1757-899X/244/1/012013

Investigation of the fatigue crack propagation behaviour in the Al alloy/Hybrid MMC Bi-layer material

AKM Asif Iqbal[1] and Yoshio Arai[2]

1 Faculty of Manufacturing Engineering, University Malaysia Pahang, 26600, Pekan, Pahang, Malaysia
2 Division of Mechanical Engineering and Science, Saitama University, 255, Shimo-okubo, Sakura-ku, Saitama 338-8570, Japan
E-mail: asifiqbal@ump.edu.my

Abstract. In this study, the fatigue crack propagation behaviour in the Al alloy-hybrid MMC bi-material system has been investigated. Three-point bending fatigue test is carried out on the Shimadzu servopulser machine. The plastic replica technique is used to observe the crack growth during cyclic loading. The crack growth rate is analyzed at different stress intensity factor range, ΔK. The experimental results showed that the crack growth decelerates in the MMC layer side and maximum crack retardation occurs on the boundary of the bi-material system. Near the boundary of the bi-material, the crack tip becomes curved, which reduces the crack growth rate in the vicinity of the boundary of the bi-material even at higher ΔK. The particle-matrix interfacial debonding, as well as particle fracture, is observed in the hybrid MMC layer during fatigue loading.

1. Introduction

Discontinuously reinforced metal matrix composites (MMCs) have attracted significant research interest in the past few decades due to their excellent mechanical properties as compared to their corresponding monolithic alloys [1]. Therefore, MMCs are now increasingly used for critical structural applications in advanced sectors. MMCs composed of a light alloy matrix, such as aluminium alloy, reinforced with ceramic particulates, have a combination of high strength, high stiffness, and low density [2]. These composites can be processed by conventional processing techniques with low cost [3, 4]. These properties allow the aluminium based MMCs as the attractive materials for use in weight-sensitive and stiffness-critical engineering components. In spite of their advantages, the expansion of Al-MMCs still causes many questions that must be resolved for a better use of these materials in the advanced sector. Many engineering components involve cyclic loading in service, therefore, fatigue degradation inevitably takes place within the matrix materials [5]. Thus, the knowledge of fatigue behaviour and resultant fracture properties of such materials become crucial for the design of the structural components. With the engineering application of MMCs, the fatigue crack growth (FCG) behaviour becomes a critical factor in design, life-prediction and reliability analysis of the components made of these materials.

The FCG behaviour in particle MMCs is very much dependent on a variety of factors, including reinforcement particle volume fraction, particle size, matrix and interfacial microstructure, the presence of inclusions and testing environment [6, 7]. Amongst these variables, particle size is one of the microstructural parameters influencing the FCG behaviour [8]. Sugimura et al. [9] have reported that increasing the volume fraction of SiC particulates in cast Al alloy stimulates higher FCG rate. Moreover, Chawla et al. [10] have investigated the effects of SiC volume fraction and particle size on

IOP Conf. Series: Materials Science and Engineering **244** (2017) 012013 doi:10.1088/1757-899X/244/1/012013

the fatigue behaviour of powder metallurgy 2080 Al alloy and found that increasing volume fraction and decreasing particle size result in an increase in fatigue resistance. Xu et al. [11] observed the retardation of FCG when crack propagates from a low volume fraction of SiC to a high volume fraction of SiC. Both crack deflection and branching decrease FCG rates. Other than the reinforcement characteristics and matrix microstructure, factors like crack closure, residual stress and elastic-plastic mismatch of matrix and reinforcement play a significant role in the FCG mechanism. Mason and Ritchie [12] observed that the crack growth resistance in the composite is superior than the monolithic alloy at low-stress intensity factor ΔK, owing to the formation of the twisted crack path, which enhanced roughness induced crack closure. In addition, Iqbal et al [13] found a higher threshold stress intensity factor range, ΔK_{th}, in the hybrid MMC as compared to whisker reinforced MMC. They explained the crack growth behaviour in terms of elastic-plastic mismatch of the reinforcement-matrix and the void nucleation in the matrix part. However, all these above phenomenon of crack profile and crack growth was explained when the crack developed in the homogeneous MMC material. The FCG in the bi-material system has been limited investigation. Few studies showed that in the case of a bi-material system, the thermal residual stress as well as the elastic mismatch of the different components of the bi-material influence the FCG rate. This mismatch of elasticity is responsible for load partitioning and elastic stress singularities that affect the value of the stress intensity factor, ΔK. Although these researches provided some guidelines of the crack growth phenomena in the bi-material system, still there is a lack of understanding of how the crack propagates in the interface region and which factors affecting the crack growth mechanism in this circumstance. Therefore, in this study, the fatigue crack propagation behaviour in an Al alloy-hybrid MMC bi-material has been investigated and the factors affecting the mechanism have been analysed.

2. Materials and experimental procedures

The bi-material used in this study is composed of Al alloy and hybrid MMC. The Al alloy-hybrid MMC bi-material and the microstructure of this material is shown in Figure 1. In the monolithic alloy part, Al alloy AC4CH is used. This alloy is also used as the base material in the hybrid MMC part. The chemical composition of the Al alloy AC4CH is shown in the Table 1. In the hybrid MMC part, the Al alloy is reinforced with 21 vol% of SiC particles and 9 vol% of Al_2O_3 fibers. In the subsequent part of this manuscript, we will call this Al alloy- hybrid MMC bi-material system as bi-material.

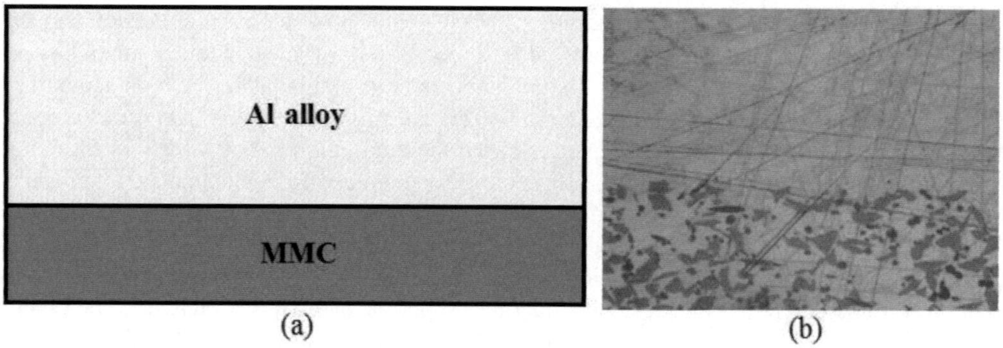

Figure 1. (a) Al alloy-hybrid MMC bi-material, (b) Microstructure of bi-materials.

Table 1. Chemical composition of AC4CH alloy, (wt%).

Si	Mg	Fe		Ti	Al
7.99	0.57	0.2 (max)		0.07	Bal.

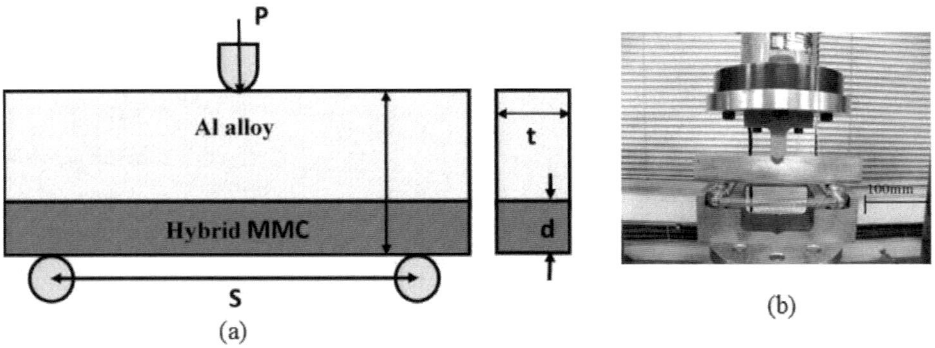

Figure 2. Specimen dimension and experimental set-up.

The material is fabricated by the squeeze casting process. Hybrid preform of SiC particles and Al_2O_3 fibers is prepared and used for the fabrication of hybrid MMC part. The preform is placed in the mould and the molten Al alloy is poured into it. The casting pressure, 100 MPa is maintained throughout the process to overcome the resistance against the flow during casting. This pressure also helps to sufficiently press the melt and fill all the open pores of the composite preforms. Finally, all the materials are heat treated by the T7 process. All the specimens are polished by using the high precision polishing machine. The polishing is carried out by using the abrasives containing 15 μm, 3 μm, 1 μm diamond particles and 0.3 μm alumina particles sequentially to remove all the machining marks and scratches from the specimen surface. To investigate the crack growth in the bi-material, 0.5 mm notch is cut in the MMC layer side. Conventional three-point bending fatigue test is carried out on the Shimadzu servopulser machine in accordance with the guideline in ASTM E647. The dimension of the specimen of the bi-material and the schematic illustration of the 3-point bending test is shown in the Figure 2. Experiments are conducted at the constant load ratio $R = 0.1$, while the frequencies vary from 1 to 20 Hz. The crack is introduced from the MMC layer side. The stress intensity factor, ΔK is increased by 5% each time the crack growth stops. The ΔK is varied depending on the longitudinal elastic modulus of the material. The Plastic replica technique is used to observe the crack growth on the surface of the specimen during cyclic loading. This technique is also used to observe the presence of the micro-cracks near the crack tip that may develop during the fatigue testing. Consequently, the FCG behaviour in the MMC layer and in the vicinity of the boundary of the bi-material is analyzed.

3. Results and discussion

The fatigue crack growth rates in the bi-material at different ΔK and at a constant load ratio, $R= 0.1$ are shown in the Figure 3 and 4. Figure 3 shows the crack growth rate in the MMC layer while Figure 4 shows the crack growth rate in the Al alloy part. The gradient of the crack growth rate is found similar in both the cases. It is observed in Figure 3 that the crack propagation slows down in the MMC layer side at higher ΔK. According to the previous study, this phenomenon of crack retardation in the MMC is due to the effect of residual stress that occurs on the specimen surface during fatigue loading [14]. However, in this research deceleration of the crack growth is also found in the Al alloy layer (Figure 4), where the residual stress is within the tensile range. Moreover, deceleration of the crack growth rate is found maximum near the boundary as shown in both the figures. According to Kang et al [15], the elastic mismatch due to the different Young's modulus of the bi-material components affects the value of the stress intensity factor (ΔK) that eventually influences the crack growth rate. Usually, an elastic-plastic interaction mismatch occurs in the composite that intensifies the crack growth rate. However, the presence of the large volume of ceramic particles creates a roughness induced crack closure that reduces the crack driving force [13]. In addition, it has been seen that when the crack approaches the boundary of the bi-material system from the higher modulus material, the ΔK value increases as compared to the monolithic material that subsequently reduces the FCG in the interface. In this research, the crack grows from the MMC layer side and penetrates to the boundary.

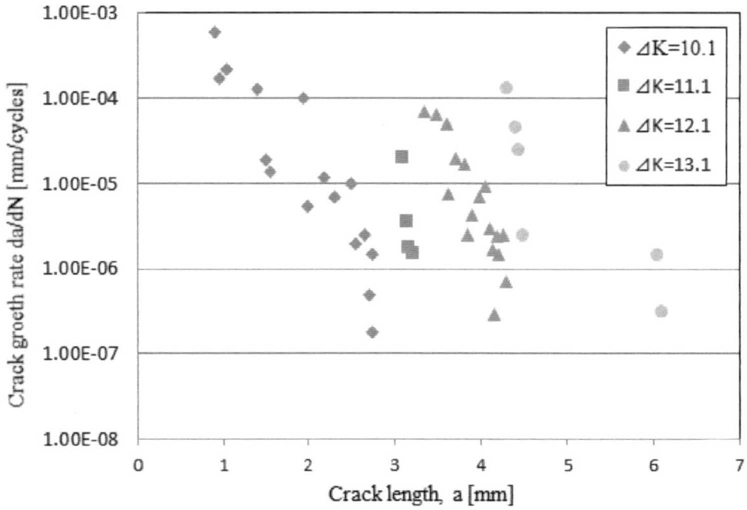

Figure 3. Crack growth rate in the hybrid MMC layer of the bi-material.

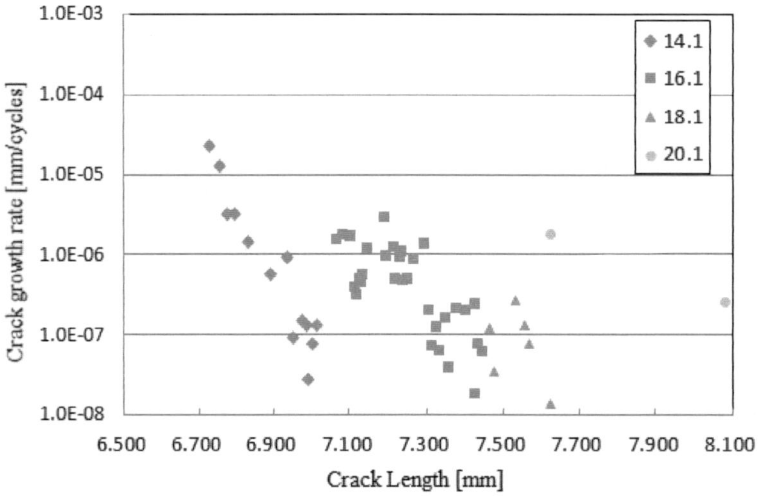

Figure 4. Crack growth rate in the Al alloy layer side of the bi-material

As both the components of the bi-material have a different elastic modulus, modulus mismatch, as well as elastic-plastic interaction mismatch, occurs at the interface. However, due to the roughness induced closure mechanism, the crack driving force is reduced, resulting in decelerating crack growth at the interface. The crack propagation behaviour on the surface of the MMC layer side near the boundary of the bi-material is shown in the Figure 5. The crack propagation direction is from the top to the bottom of the figures. Figure 5(a) shows that the crack is developed at the interface of the ceramic particle and Al matrix. This crack moves to the base material in subsequent fatigue cycle as shown in Figure 5(b). Another crack progression is shown just before the ceramic particles in Figure 5(c). This crack propagates and broke the SiC particles as shown in the circle in Figure 5d. Finally, the crack moves through the base material and stops. This observation indicates that the crack due to fatigue initiates in the particle-matrix interface and propagates through the interface as well as the base material. The interface debonding and the particle fracture occurs as the fatigue cycle increases. The crack stops at the particle or at the base material if the corresponding ΔK is low.

The fracture surface near the boundary of the bi-material specimen is analyzed to investigate the nature of the crack tip, and the surface is shown in the Figure 6. The fracture surface is found dark near the boundary as shown by the bracket in the figure 6(a) but bright at the center of the specimen. This is due to the difference in crack propagation speed at the boundary and in the center of the specimen. It is found that the fracture occurs in different ways in different places of the specimen. Tearing is observed at the center of the specimen while fatigue failure is observed in the surface. The schematic illustration shown in Figure 6(b) indicates the different nature of the fracture in the bi-material. During the experiment, the crack length is measured on the surface of the specimen as well as beneath the surface. The crack length measured on the surface is found larger than the crack length found in the specimen. This indicates that the crack tip moves far on the surface compared to that of the crack tip inside the material. Therefore, a curved crack tip (corner crack) is developed in the specimen. Due to the nature of the curved crack tip, the deceleration of the crack growth occurs in the vicinity of the boundary of the Al alloy-hybrid MMC bi-material system.

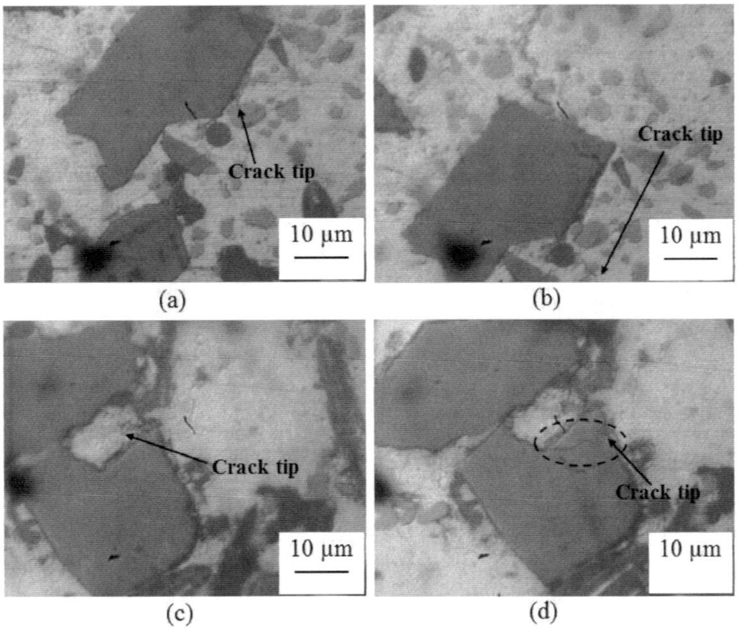

Figure 5. Fatigue crack propagation in the MMC layer side near the boundary of the bi-material.

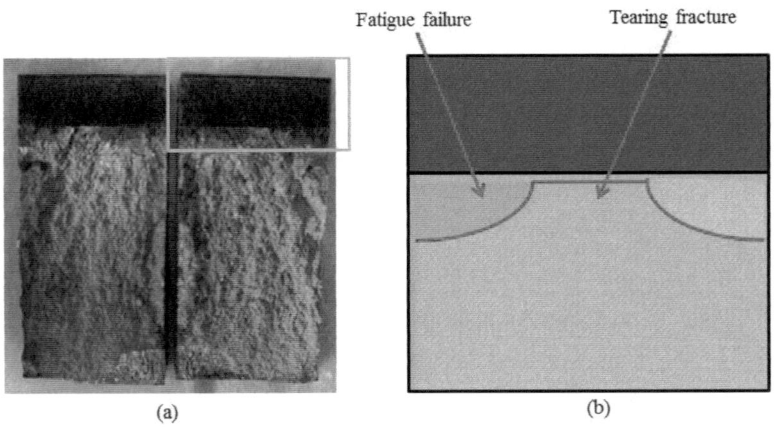

Figure 6. (a) Fracture surface of the bi-material, (b) Schematic illustration shows different types of fracture in bi-materia.l

4. Conclusions
In this research, the fatigue crack propagation behaviour in the Al alloy-hybrid MMC bi-material system has been investigated. The results highlights that the crack propagation slows down in the MMC layer side and maximum crack retardation occurs on the boundary of the bi-material system. The crack in the MMC layer initiates in the particle-matrix interface and propagates through the base material. The particle-matrix interfacial debonding, as well as particle fracture, is observed during fatigue loading. Near the boundary of the bi-material, the crack tip is found curved due to the difference of crack growth rate on the surface and inside the specimen. Therefore, a corner crack is generated. Due to this corner crack, the deceleration of the crack growth occurs in the vicinity of the boundary of the bi-material even at higher ΔK.

5. References
[1] Miracle DB 2005. Metal matrix composites – from science to technological significance. Compos Sci Technol 65, 2526–2540
[2] Mkaddem A., Mansori M. El 2009. On fatigue crack growth mechanisms of MMC: Reflection on analysis of 'multi surface initiations'. Mater Des 30, 3518–3524
[3] Chen ZZ, Tokaji K 2004. Effects of particle size on fatigue crack initiation and small crack growth in SiC particulate-reinforced aluminium alloy composites. Mater Lett 58, 2314–21
[4] Yang X, Liu CR. Investigating machining impact on fatigue variance: a machining quality perspective. J Manuf Sci Eng 2005;127(3):492–502
[5] Paepegem VW, Degrieck J 2002. A new approach of residual stiffness and strength for fatigue of fibre-reinforced composites. Int J Fatigue 24, 747–762
[6] Chawla N, Ganesh VV, Wunsh B 2004. Three-dimensional (3D) microstructure visualization and finite element modeling of the mechanical behavior of SiC particle reinforced aluminum composites. Script Material, 51, 161-165
[7] Chawla N, Andres C, Davis LC, Jones JW, Allison JE 2000. The interactive role of inclusion and SiC reinforcement on the high-cycle fatigue resistance of particle reinforced metal matrix composites. Metall Mater Trans A 31, 951-957
[8] Llorca J, Ruiz J, Healy JC, Elices M, Beevers CJ 1994. Fatigue crack propagation in salt water, air and high vacuum in a spray-formed particulate-reinforced metal matrix composite. Mater Sci Eng A 185, 1-15
[9] Sugimura Y, Suresh S 1992. Effects of SiC contents on fatigue crack growth in aluminum alloys reinforced with SiC particles. Metall Mater Trans A, 23, 2231-2242
[10] Chawla N, Andres C, Jones JW, Allison JE 1998. Effect of SiC volume fraction and particle size on the fatigue resistance of a 2080 Al/SiCp composite. Metall Mater Trans A, 29, 2843-2854
[11] Xu FM, Zhu SJ, Zhao J, Qi M, Wang FG, Li SX, Wang ZG 2003. Fatigue crack growth in SiC particulates reinforced Al matrix graded composite. Mater Sci Eng A, 360, 191-496
[12] Mason JJ, Ritchie RO 1997. Fatigue crack growth resistance in SiC particulate and whisker reinforced P/M 2124 aluminum matrix composites. Mater Sci Eng A, 231, 170−182
[13] Iqbal AA, Arai Y, Araki W 2014. Fatigue crack growth mechanism in cast hybrid metal matrix composite reinforced with SiC particles and Al2O3 whiskers. Trans. Nonferrous Met. Soc. China 24, 1−13
[14] M.T. Milan, P. Bowen 2003. Experimental and predicted fatigue crack growth resistance in Al2124/Al2124+35%SiC biomaterial. Int J Fatigue 25, 649-659
[15] Kang KJ, Song JH, Earmme YY 1990. Fatigue crack growth and closure behaviour through a compressive residual stress field. Fatigue Fract Eng Mater Struct 13, 1–13

IOP Conf. Series: Materials Science and Engineering **244** (2017) 012014 doi:10.1088/1757-899X/244/1/012014

Adsorption and desorption behavior of CO on S passivated 4H-SiC by electric field

Zhengxin Yan, Anhong Hou, Yongsheng Wang and Shaohua Zhang

College of Science, Xi'an University of science and technology, Xi'an, China

Abstract. Adsorption energy and partial density of states of S at three typical positions on 4H-SiC (001) surface were calculated which based on the density functional theory (DFT).The results show that S adsorption is the most stable in the bridge position and unstable in the hollow. The density of states near the Fermi level can be effectively reduced by using the S atom on the surface of 4H-SiC. The adsorption and desorption of CO on the surface of S passivated 4H-SiC at different applied electric fields were calculated by selecting hollow of S atom adsorption. It turned out that the positive electric field is not conducive to the adsorption of CO and the reverse is favorable for adsorption, but when the forward electric field F is greater than 2eV/ Å /e the desorption behavior of S atom is more obvious. Because of the S atom is positively charged which made the Si-C bond length increases rapidly under the action of electric field.

1. Introduction

There are some advantages in SiC including The band gap、saturated electron drift velocity、electron mobility and high breakdown voltage, so the fabrication of semiconductor materials have been widely used in high temperature、high frequency and anti radiation and other extreme environments. One of the most important problems in semiconductor devices and circuits is how to reduce the surface dangling bonds or reduce the density of states on the surface. As a result, many people have studied the use of O、H、 N、P and other elements to inactivate SiC power devices, while S passivation technology is also widely used in GaAs、ZnP、Ge and other semiconductor materials.[1-4] Using S passivation technique showed that S blunt Having a low surface state exhibits a better device performance . [5-6]

In general, the electric field force has a significant effect on the properties of the electron in the low dimensional system, including one-dimensional quantum wire [7], carbon nanotube [8, 9] and graphene. [10] Many phenomena have been studied in the tight binding model and density functional theory calculations.[11] The results show that the band and the density of states are very sensitive to the direction and intensity of the electric field. The adsorption and desorption of N2 on Fe (111) surface by electric field F were studied by the first principles method. [12] Meanwhile the effect of F on the surface of Pt (111) [13] and the adsorption and desorption of CO on the surface of graphene were studied. [14] The results show that F has changed the orbit of electron, so some new physical phenomena have appeared.

We are interested in the effect of S atoms adsorbed on the 4H-SiC (001) surface on the surface state density and the influence of the electric field force on the adsorption and desorption behavior of CO on the 4H-SiC surface under S passivation.

IOP Conf. Series: Materials Science and Engineering **244** (2017) 012014 doi:10.1088/1757-899X/244/1/012014

2. Calculation method

In this paper we use the Castep module in the Material software Studio and the electronic exchange correlation part is described by the PW91 standard exchange correlation function under the generalized gradient approximation (GGA). The integral of Brillouin zone is selected by using the special K point grid method of Monkhorst-Pack and the grid size of K is set to 2 x2 x2. The cut off energy can be set to 380eV. The convergence criteria of the selected energy is 1 x10^{-5}eV/atom.The size of the selected super crystal package is 3 x3 x1 meanwhile Si-C bond length is1.88 Å 、 C-O bond length is 1.13 Å and Si-S bond length is 2.1 Å .In order to make the interaction between the surfaces do not interfere with the calculation the thickness of the vacuum layer along the Z direction is 13 Å .The two layers of C and Si atoms in the bottom layer are fixed, so that the dipole moment of the bottom atom has the least influence on the upper layer atoms and the adsorbed atoms, while made the amount of calculation reduced.
When a CO adsorbed on the 4H-SiC (001) surface the sulfur atoms are adsorbed on the positions of the three typical high symmetry as shown in figure Figure 1.

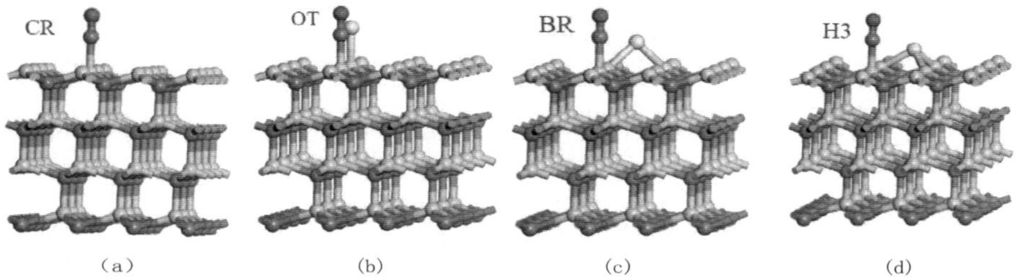

Figure 1. (a) 4H-SiC (001) surface adsorb CO but from (b) to (d) meanwhile adsorb S atom on three adsorption site of his surface. The on-top site (OT) is located on the top of a surface Si atom, the bridge site (BR) is a two-fold coordinated site between two surface Si atom, the three-fold hollow site (H3) is above a carbon of the third bilayer.
The adsorption energy of each sulfur atom is calculated by the following formula:

$$E^{s}_{ads} = -(E_{s/co/slab} - E_{co/slab} - NE_{s})/N \quad (1)$$

Here N is the number of S atoms adsorbed on 4H-SiC (001) surface, $E_{s/co/slab}$、 $E_{co/slab}$、 E_{s} are the total energy of the surface, total surface energy without sulfur atoms and the total energy of the sulfur atom.
The most stable adsorption sites and the most unstable adsorption sites of S atoms were found from the calculation of adsorption energy as and the density of States. The passivation effect of S atoms on 4H-SiC (001) and the influence of applied electric field on the adsorption of CO on the adsorption of S atoms in the most unstable positions were studied. Here we choose the hollow to Adsorb sulfur atoms. The direction of the applied electric field is along the Z axis (positive electric field along the positive direction of the Z axis and negative electric field in the opposite direction).electric field intensity are-2.5、-2、-1.5、-1、1、1.5、2、2.25、2.5 eV/ Å /e. In this model the length of the electric field direction is 20.89 Å , so we add the voltage up to 52V.The adsorption energy of CO under the action of electric field force F is determined by the following formula:

$$E_{ad}(F) = E_{s/co/slab}(F) - E_{s/slab}(F) - E) \quad (2)$$

$E_{s/co/slab}(F)$、 $E_{s/slab}(F)$ and $E_{co}(F)$ indicate that the total energy of the CO and S atoms adsorbed on the external electric field, the total surface energy of without CO molecule and the total energy of CO under the corresponding electric field .

IOP Conf. Series: Materials Science and Engineering **244** (2017) 012014 doi:10.1088/1757-899X/244/1/012014

3. Results and discussion

The adsorption energies of S atoms in three typical positions were calculated by 7.05, 7.23, and 6.98 eV, so the best adsorption position of S was found at the bridge site, followed by the top position and most unstable was located in hollow. Figure 2 shows the density of the states of the atom, which is not adsorbed on the S atom and the adsorption of S atoms in the typical positions. It can be concluded from the state density diagram that the adsorption of S atoms decreases the density of states near the Fermi level of SiC. As well as, The density of states in the bridge decreases the most while the decrease in hollow is the least.

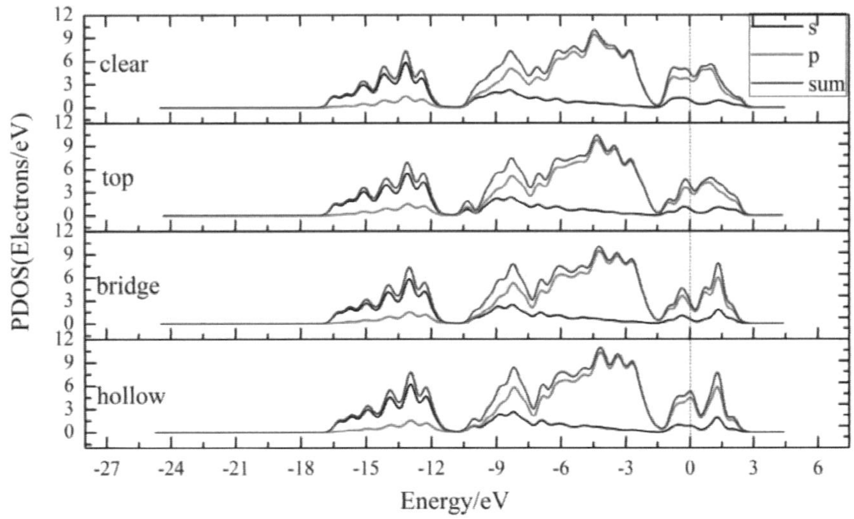

Figure 2. The density of states is no adsorption of S atoms and S atoms adsorbed at three typical positions.

Due to the distribution of the 4H-SiC surface atoms and the internal atoms differ in structure and surface energy minimization constraint makes the surface atoms reconstruct, resulting in differences in the nature of the surface atoms and inside of atoms in physical and chemical aspects of relatively large. Choosing a suitable adatom to adsorb a layer of passivation on the surface of 4H-SiC can restore the geometry of the surface atoms to the structure of the internal atoms. The S atom is chosen as the passivation atom, mainly because the S atom is closest to the corresponding parameters of the Si atom in terms of the covalent radius, atomic radius, bond length and bond angle, thus meeting the requirements of Valence-Mending. The adsorbed S atoms saturate the surface of the partial dangling bonds and the surface atoms form stable valence bonds, which reduces the dimer bonds and stress bonds due to the introduction of the reconstruction. This explains why S atom passivation reduces the surface density of States.

It is concluded that the sulfur atom has a surface passivation effect on 4H-SiC and the density of the surface is reduced obviously. The model of the S atom adsorbed on the hollow position is obtained after the applied electric field as shown in Figure 3.It can be found that the adsorption of CO enhance with the increase of electric field in the negative electric field. With the increase of the electric field in positive direction the adsorption of CO is weaken and there is a trend of desorption before F=2.But the interesting thing is that when the electric field is greater than 2 we can find that the S atoms change greatly. When F=2.5 the S atoms change more than CO and the maximum length of Si-S is up to 4.829 Å .This is mainly due to the ground state of Si-S and energy spectrum near the excited state, along with the increase of electric field strength S atomic electrons being excited state as the field continues to increase was finally out then S electronic excited atoms form a positively charged particle. The positive charge moves in the direction of the electric field and the bond length is elongated rapidly when the electric field is very strong. This is why the length of the S-Si bond grows suddenly.

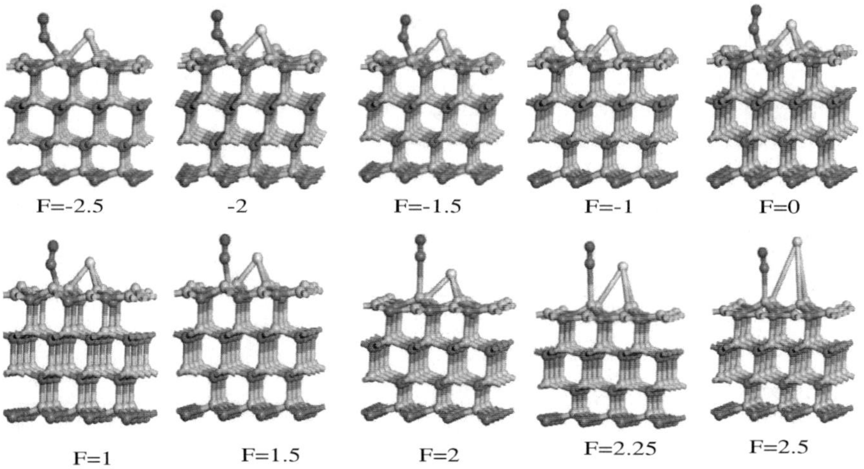

Figure 3. Optimal structure under different electric fields.

Figure 4 is the adsorption energy of CO under different electric field, Since the electric field strength reaches 2.25, the structural optimization can not be convergent, so the adsorption energy is not listed here. We can see that the change trend of $E_{ad}(F)$ is that the reverse electric field is beneficial to strengthen the adsorption, and the positive electric field is weakened. This is consistent with the previous analysis of the optimized structure, so a high electric field force can be used to sharpen the sensitive material to be repeated .There is no approximate linear relationship between the adsorption energy and the applied electric field and the adsorption energy is approximately linear with the applied electric field. [14]This linear relationship can be explained by first-order stark effect. There is a step in our model which is related to the adsorption of S atoms. The S atom has a passivation on the surface atoms and has an effect on the adsorption of CO. It can be seen that S enhanced the adsorption of CO on the negative electric field.

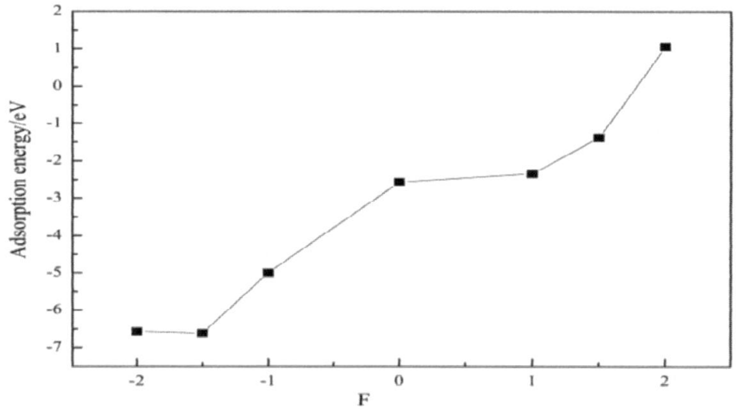

Figure 4. Adsorption energy of CO under different electric fields.

It can be found from the Table 1 that the bond length of CO increases with the increase of the electric field when the electric field is added along the negative direction, and the bond length decreases with the increase of the electric field strength when the electric field of forward is applied to 1.5.Because the C atom is positively charged and the O atom is negatively charged, the electric potential decreases along the direction of the electric field, which causes the decrease of the CO molecular distance. The same charge mutual repulsion and mutually opposite charges attract is another reason. When the electric field is in the negative direction, the CO molecule is a polar molecule, so the potential decreases along the direction of the electric field. When the electric field intensity changes from 1.5 to

2.25, the bond length of CO increases, which is mainly affected by the adsorption of S atoms.∠O1-C37—Si29 bond angle increases with the increase of electric field strength. One of the three Si atoms bonded to the S atom is a Si-S bond away from the CO molecule with a change in the electric field close to a straight line only when the F is greater than 2 suddenly changes greatly, which is related to the excitation of the electrons of the S atom mentioned above. The change of C-Si bond length is the same as that of CO.

Table 1. Change of bond length bond angle under different electric fields.

	F=-2.5	F=-2	-1.5	-1	0	1	1.5	2	2.25	2.5
C37—O1	1.221	1.219	1.21	1.214	1.193	1.172	1.151	1.168	1.2	1.159
C37—Si29	2.393	2.288	2.062	2.1	1.993	1.983	1.969	2.935	2.999	2.49
∠O1-C37—Si29	133.59	135.314	136.203	138.073	138.924	139.799	163.382	178.432	177.088	172.472
Si17-S	2.503	2.47	2.45	2.413	2.401	2.346	2.343	2.381	3.156	4.587

In Figure 5, the difference density of charge is given under different electric fields, where the red represents the accumlation of electrons and the blue represents the absence of electrons.CO and 4H-SiC belong to a covalent bond because of the aggregation of electrons in a joint rather than an atom. Since C、O、Si、S atoms have different electronegativity, the electron from the Si-C bond is more likely to tend to the C bond and the C-O bond is more inclined to the O bond. It can be seen from the figure that with the increase of the negative electric field the electrons at the C end will increase and the C will decrease when the forward electric field increases. The absence of electrons on the side of the S atom close to the CO molecule has an effect on the bond length and adsorption of CO. With the increase of the electric field the electrons decrease more from the S atoms. When the F is greater than 2 generous electrons of the S atom are reduced so that we can think of the positive charge of the S atom, which is consistent with the previous analysis.

Figure 5. Differential charge density at different electric fields; The red region means electronic accumulation, while the blue region means electronic loss.

4. Conclusion

Based on the first principles of density functional theory, three typical positions of S adsorbed on 4H-SiC (001) surface that a CO have been adsorbed were studied .The density of States and adsorption energy were calculated. the results show that the S atom in the most stable bridge adsorbed on hollow adsorption of the most unstable, and the surface S passivation 4H-SiC effectively reduces the density of states near the Fermi level. The adsorption and desorption of S on 4H-SiC surface under different external electric field were studied. It was found that the positive electric field was unfavorable to CO adsorption, and the reverse electric field was favorable for the adsorption of CO. However, when the positive electric field F is greater than 2eV/ Å /e, the desorption behavior of S atoms is more obvious, mainly due to the increase of the electric field strength will cause the electron transfer of S atoms. At this point, the S atom is positively charged, and the Si-C bond length increases rapidly under the action of electric field.

5. References

[1] Ruff M, Mitlehner H and Helbig R 1994 IEEE Trans. Electron Devices **41** 1040
[2] Nakamura D, Gunjishima I, Yamaguchi S, ItoT, Okamoto A, Kondo H, Onda S and Takatori K 2004 Nature. **430** 1009
[3] Melinon P, Masenelli B, Tournus F and Perez A 2007 Nat. Mater. **6** 479
[4] Presser V and Nickel K G 2008 Crit.Rev. Solid State Mater. Sci.**33** 1
[5] Wenbo Li, Ling Li, Fangfang Wang, China.Phys.B.**26**(2017) 037104
[6] He Xiao-Min, Chen Zhi-Ming , MAterials Science Forum. **858** (2016) 1885-188
[7] J.Singh,Physics of Semiconductors and Their Heterostructures, McGraw-Hill, New York, 1993
[8] J. O'Keeffe, C.Y. Wei, K.J. Cho, Appl. Phys. Lett. **80** (2002) 676
[9] Y. Li, S.V. Rotkin, U. Ravaioli, Nano. Lett. **3** (2003) 183
[10] Z.M. Ao, W.T. Zheng, Q. Jiang, Nanotechnology **19** (2008) 275710
[11] C.L. Lu, C.P. Chang, Y.C. Huang, R.B. Chen, M.L. Lin, Phys. Rev. B **73** (2006)144427
[12] D. Tomanek, H. Kreuzer, J. Block, Surf. Sci. **157** (1985) L315
[13] M.P. Hyman, J.W. Medlin, J. Phys. Chem. B **109** (2005) 6304
[14] Ao Z M, Li S, Jiang Q.graphene[J]. Solid State Communications. **150** 13(2010)680-683

IOP Conf. Series: Materials Science and Engineering **244** (2017) 012015 doi:10.1088/1757-899X/244/1/012015

Analyzing of a 2-D Magnet Array with Hexagon Magnet based on superposition

Guangdou Liu [1], Xin Zhang[1], Xingping Xu[1] and Xiaoyi Liu[2]

1 College of Mechanical and Electronic Engineering,
China University of Petroleum, Qingdao 266580, China
2 College of arts, China University of Petroleum, Qingdao 266580, China
Email:gdliu@upc.edu.cn,zxin0927@163.com,xuxp@upc.edu.cn,wdliuxiaoyi@126.com

Abstract. In this paper, a 2D Halbach array with the hexagon permanent magnets is introduced. The expression of magnetization intensity is derived by superposition principle and Fourier series. The expression of magnetic flux density is derived by magnetic scalar potential and method of separation of variables. The assumption of the relative permeability of magnet equals 1.0 is needless because of the compact structure. The accuracy of the expression is verified by comparing with the finite element model. The expression can be directly used in real time control by selecting the proper iteration steps for the coefficients.

1. Introduction

According to the characteristics of Halabch arry[1], the magnetic field is significantly enhanced on one side and significantly weakened on the other side. The magnetic field of the Halbach array has good sinusoidal distribution. As for 2D Halbach array, it can be used in the magnetically levitated planar motor that can realize non-contact and non-abrasion plane motion [2]. Comparing with the conventional planar motion mechanism comprised by two linear motors, there is no supporting guide, and the motor with simple structure can realize high speed and acceleration motion. It has extensive application perspectives in high precision lithographic machines and so on [3, 4 and 5].
The different shapes except rectangular of magnets are applied to the 2D Halbach array in order to obtain higher magnetic flux density, such as trapezoid, triangle, cone, hexagon and so on [6,7 and 8]. The disadvantage is that the expressions of magnetic flux density deriving by analytical method are complicated. It is difficult to obtain the real time control mode. In this paper, take the 2D permanent magnet array using hexagon magnets for example [9]. The quasi-analytical expression of magnetic flux density is derived by using the superposition principle. So the real time control mode can be easily obtained by the quasi-analytical expression. The method can be applied to analysis of the permanent magnet array with the non-rectangular magnets.

2. Analysis of magnet flux density

Figure 1 shows the permanent magnet array with a top view and a cut-view. In figure 1, the arrow denotes the magnetization direction of the magnets from S-pole to N-pole. S means the direction towards the paper and N means out of paper. The hexagon magnets are in horizontal directions. τ is the pole pitch, τ_m is the length of the side of the magnets which are magnetized in z-direction, τ_d is the length of the side of the hexagon magnets.
The permanent magnets can be modelled as an infinitely large magnet array with Fourier series by neglecting the end-effects [10]. The residual magnetization vector can be expressed as

$$\vec{M} = M_x\hat{x} + M_y\hat{y} + M_z\hat{z} \tag{1}$$

The component of M_x can be seen as the distribution of hexagon magnets because the magnetization direction. In this paper, the hexagon magnet can be seen as a superposition of an infinite number of small rectangular segments. The derivation of M_x of the hexagon magnets is based on the superposition principle. Figure 2 shows the M_x projection distribution of one rectangle segment in the hexagon magnets, including the x-direction and y-direction.

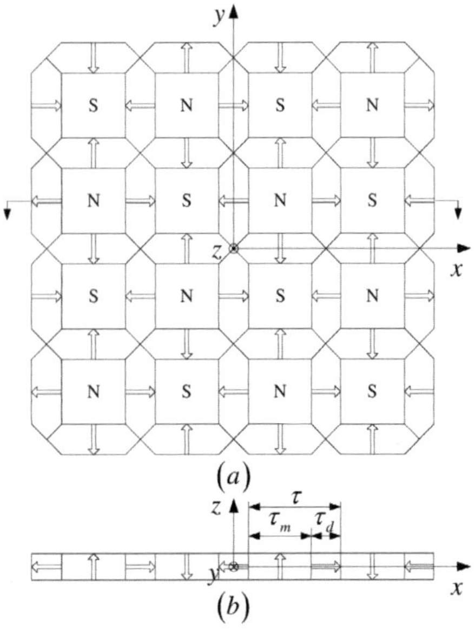

Figure 1. The permanent magnet array (a) Top view. (b) Cut-view.

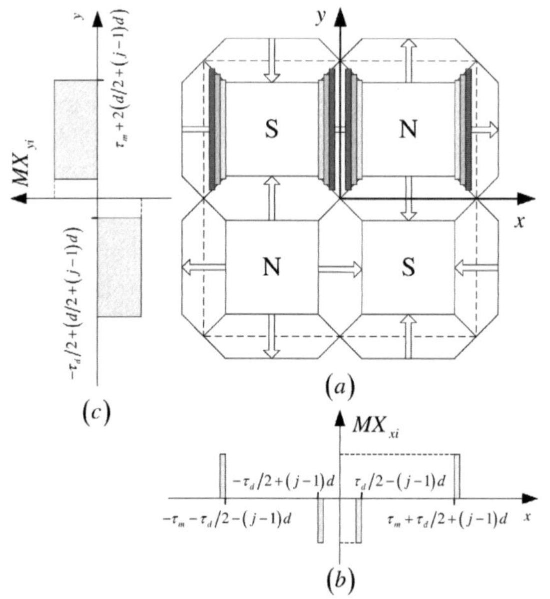

Figure 2. M_x projection distribution of one segment. (a) Magnet array. (b) x-direction. (c) y-direction.

The projection distribution is derived by using Fourier series. Then the expression of M_{xj} is

$$M_{xj} = MX_{xj} \cdot MX_{yj} \cdot M = -M\sum_{k=1}^{\infty}\sum_{l=1}^{\infty} a_j(k)b_j(l)\cos(k\omega x)\sin(l\omega y) \tag{2}$$

where $M = \frac{B_r}{\mu_0}, \omega = \frac{\pi}{\tau}$, B_r is the remanent magnetization of the permanent magnet, μ_0 is the vacuum permeability, j represents the sequence number of the segments, k and l are the harmonic numbers for the x- and y-direction, respectively, $a_j(k)$and $b_j(l)$ are projection distribution coefficients,

$$a_j(k) = \frac{8}{k\pi}\sin\frac{k\pi}{2}\sin\frac{k\omega(\tau_m + (2j-1)d)}{2}\sin\frac{k\omega d}{2} \tag{3}$$

$$b_j(l) = \frac{4}{l\pi}\sin\frac{l\pi}{2}\sin\frac{l\omega(\tau_m + (2j-1)d)}{2} \tag{4}$$

If the hexagon magnet is divided into n segments, the residual magnetization M_x can be expressed

$$M_x = \sum_{j=1}^{n} M_{xj} = -\sum_{k=1}^{\infty}\sum_{l=1}^{\infty}\left(\sum_{j=1}^{n} a_j(k)b_j(l)\right)\cos(k\omega x)\sin(l\omega y) \tag{5}$$

The other component of $\vec{M}$ can be obtained by using the same approach. The function of M_y can be established by interchanging the variables x and y in the function of M_x. The expression of the residual magnetization vector of this array is

IOP Conf. Series: Materials Science and Engineering **244** (2017) 012015 doi:10.1088/1757-899X/244/1/012015

$$\vec{M}=\frac{B_r}{\mu_0}\sum_{k=1}^{\infty}\sum_{l=1}^{\infty}\begin{bmatrix}-R_x\cos(k\omega x)\sin(l\omega y)\\-R_y\sin(k\omega x)\cos(l\omega y)\\R_z\sin(k\omega x)\sin(l\omega y)\end{bmatrix} \tag{6}$$

where,

$$c(k)=\frac{4}{k\pi}\sin\frac{k\pi}{2}\sin\frac{k\omega\tau_m}{2} \tag{7}$$

$$R_x=\sum_{j=1}^{n}a_j(k)b_j(l) \tag{8}$$

$$R_y=\sum_{j=1}^{n}a_j(l)b_j(k) \tag{9}$$

$$R_z=c(k)c(l) \tag{10}$$

In order to derive the magnetic flux density distribution of the magnet array, the 3D space is divided into three regions. The cross section is shown in figure 3. Regions 1 and 3 are in air. The permanent magnets in region 2 are located in between m_b and m_t. There is no air in the permanent magnet array because of the compact structure. So the assumption of air in the magnet array is not existed. The scalar magnetic potential equation is used to this problem because there is no current.

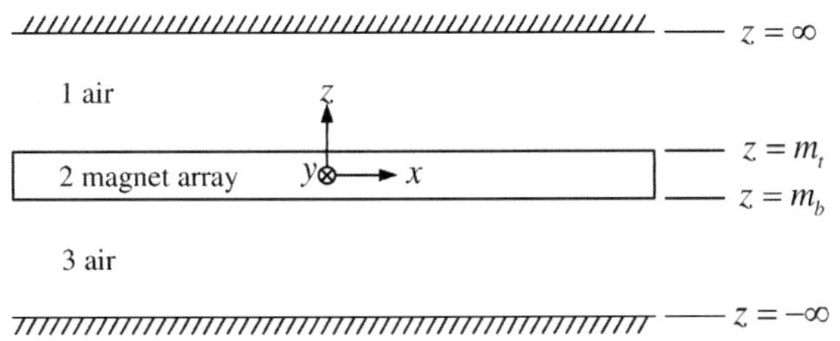

Fig. 3. Space divided into three regions.

The magnetic field equations and boundary conditions are listed in Table 1 and Table 2, respectively.

Table 1. Magnetic field equations.

Region1	Region2	Region3
$\vec{H}_1=-\nabla\Psi_1$	$\vec{H}_2=-\nabla\Psi_2$	$\vec{H}_3=-\nabla\Psi_3$
$\nabla\times\vec{H}_1=0$	$\nabla\times\vec{H}_2=0$	$\nabla\times\vec{H}_3=0$
$\vec{B}_1=\mu_0\vec{H}_1$	$\vec{B}_2=\mu_0\mu_r\vec{H}_2+\mu_0\vec{M}$	$\vec{B}_3=\mu_0\vec{H}_3$
$\nabla\cdot\vec{B}_1=0$	$\nabla\cdot\vec{B}_2=0$	$\nabla\cdot\vec{B}_3=0$
$\nabla^2\Psi_1=0$	$\nabla^2\Psi_2=\nabla\cdot\vec{M}/\mu_r$	$\nabla^2\Psi_3=0$

Table 2. Boundary conditions.

z=∞	z=m_t	z=m_b	z=-∞
Ψ_1=0	$H_{1x}=H_{2x}$ $H_{1y}=H_{2y}$ $B_{1z}=B_{2z}$	$H_{2x}=H_{3x}$ $H_{2y}=H_{3y}$ $B_{2z}=B_{3z}$	Ψ_3=0

In the table above, $\vec{B}_1$ is the magnetic flux density distribution, $\vec{H}_1$ is the magnetic field strength, Ψ_1 is the scalar potential, μ_r is the relative permeability of the magnets, m_t, m_b is the height of the magnet array.
The method of separation of variables is used to solve the Laplace equations. A solution of the scalar potential is substituted.

$$\Psi = \sum_{k=1}^{\infty}\sum_{l=1}^{\infty} Z(z)\sin(k\omega x)\sin(l\omega y) \tag{11}$$

In regions 1 and 3, by solving the function the expression is

$$Z(z) = K_1\exp(-\lambda z) + K_3\exp(\lambda z) \tag{12}$$

$$\lambda = \omega\left(k^2 + l^2\right)^{1/2} \tag{13}$$

where K_1 and K_3 are constants. Because of the boundary conditions (zero scalar potential for $z = \pm\infty$)

$$\Psi_1 = \sum_{k=1}^{\infty}\sum_{l=1}^{\infty} K_1\exp(-\lambda z)\sin(k\omega x)\sin(l\omega y) \tag{14}$$

$$\Psi_3 = \sum_{k=1}^{\infty}\sum_{l=1}^{\infty} K_3\exp(\lambda z)\sin(k\omega x)\sin(l\omega y) \tag{15}$$

In region 2, a nonhomogeneous differential equation is obtained. The solution is

$$Z_2(z) = K_2\exp(-\lambda z) + T_2\exp(\lambda z) - C \tag{16}$$

$$C = \frac{\omega B_r}{\lambda^2\mu_r\mu_0}\left(kR_x + lR_y\right) \tag{17}$$

where K_2 and T_2 are constants.
The above constants can be calculated with the boundary conditions. The expression for the magnetic flux density distribution in region 3 is

$$\vec{B}_3 = -\mu_0\sum_{k=1}^{\infty}\sum_{l=1}^{\infty} K_3\exp(\lambda z)\begin{bmatrix} k\omega\cos(k\omega x)\sin(l\omega y) \\ l\omega\sin(k\omega x)\cos(l\omega y) \\ \lambda\sin(k\omega x)\sin(l\omega y) \end{bmatrix} \tag{18}$$

where

$$K_3 = -B_r\frac{\exp(-\lambda m_b)\left(\exp(-\lambda m_t) - \exp(-\lambda m_b)\right)\left(d_2kR_x + d_2lR_y + d_1\left(k^2+l^2\right)^{1/2}R_z\right)}{\mu_0\omega\left(k^2+l^2\right)\left[(\mu_r+1)^2\exp(-2\lambda m_b) - (\mu_r-1)^2\exp(-2\lambda m_t)\right]} \tag{19}$$

$$d_1 = \mu_r\left(\exp(-\lambda m_t) - \exp(-\lambda m_b)\right) + \left(\exp(-\lambda m_t) + \exp(-\lambda m_b)\right) \tag{20}$$

$$d_2 = \left(\exp(-\lambda m_t) - \exp(-\lambda m_b)\right) + \mu_r\left(\exp(-\lambda m_t) + \exp(-\lambda m_b)\right) \tag{21}$$

The magnetic flux density at 4 mm below the magnet array is shown in figure 4. τ_m=17mm, B_r=1.24T, n=4, τ=25mm, the thickness of magnet array is 7mm. The magnetic flux density calculated by 3-D finite element model is shown in figure 4(a). The error between the quasi-analytical model and finite element model is shown in figure 4(b). The maximum difference value between the quasi-analytical model and finite element model is less than 0.015 T, which is 2.64% of the peak value of the finite element model.

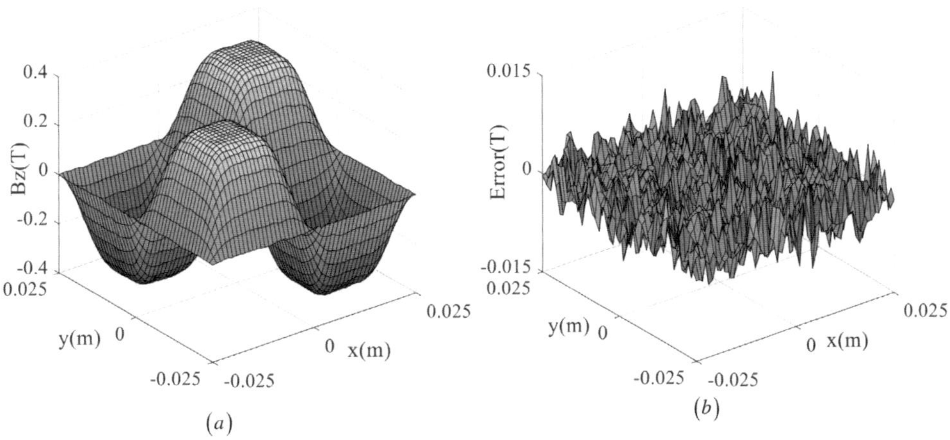

Figure 4. Magnetic flux density, (a) Calculated by the finite element model, (b) Error between the quasi-analytical model and the finite element model.

3. Conclusions

The 2D Halbach array with compact structure by using the hexagon permanent magnets is introduced. The quasi-analytical expression of magnetic flux density is derived by the superposition principle. The accuracy is verified by comparing with finite element model. The real time control mode can be easily obtained by the quasi-analytical expression. The method can be applied in the design of permanent magnet array with the non-rectangular magnets of planar motors.

4. References

[1] Klaus H 1985 *J. Appl. Phys.* **57** 3605
[2] J W Jansen, J P C Smeets, T T Overboom, J M M Rovers and E A Lomonova 2014. IEEE *Trans. Magn.* **50** 8500404
[3] J C Compter 2004 *Precision Eng.* **28** 171
[4] Y Ueda, H Ohsaki 2008 *IEEE Trans. Magn.* **44** 609
[5] L Zhang, B Kou, H Zhang, S Guo 2012 *IEEE Trans. Magn.* **48** 4658
[6] H S Cho, H K Jung 2002 *IEEE Trans. Energy Conver.* **17** 492
[7] Y Zhang, Z Yang, M Yu, K Lu, Y Ye, and X Liu 2011 *IEEE Trans. Magn.* **47** 3236
[8] J R Peng, Y F Zhou 2013 *IEEE Trans. Magn.***49** 618
[9] G Liu, Y Zhou, R Zhou, W Ming and L Huo 2013 *APPL Comput Electrom* **28** 976
[10] H S Cho, C Hn Im and H K Jung 2001 *IEEE Trans. Magn.* **37** 3966

Acknowledgments

Research was supported by the Natural Foundation of Shandong Province (ZR2015EQ015) and the Fundamental Research Funds for the Central Universities (15CX02031A).

IOP Conf. Series: Materials Science and Engineering **244** (2017) 012016 doi:10.1088/1757-899X/244/1/012016

Constitutive modelling of elastomer/graphene platelet nanocomposites

Amir A. Abdelsalam[1], Sherif Araby[2,3], M. A. Hassan[1,4*] and A. A. El-Moneim[1,5]

1 Materials Science and Engineering Department, Egypt-Japan University of Science and Technology, Alexandria, Egypt
2 Department of Mechanical Engineering, Faculty of Engineering, Benha University, Egypt
3 School of Engineering, University of South Australia, SA 5095, Australia
4 Mechanical Engineering Department, Faculty of Engineering Assiut University, Assiut 71516, Egypt
5 Physical Chemistry Dept., National Research Center, El-Dokki-Giza, Egypt
Email: *mohsen.khozami@ejust.edu.eg

Abstract. Elastomers are used in a wide variety of structural and engineering applications. They exhibit a nonlinear elastic stress-strain behaviour known as hyperelasticity which is generally described by hyperelastic strain energy functions. The question raised in the current study was; which model can accurately describe and predict the actual behaviour of the elastomer nanocomposites. The tensile data were used to fit the various elastomeric material models available in MSC.MARC finite element analysis package. The relative percentage error was calculated to determine the goodness of fit in order to select the best model. Numerical results showed that the third order deformation model was the best among the various material models since giving a maximum relative error of fit was 2.7% at small and large strains. To verify the effectiveness of third order deformation model, FE simulations for tensile test was carried out. The results showed that the third order model is sufficiently enough to regenerate the experimental data for uniaxial test and efficiently capture the hyperelastic behavior as good as the experiments.

1. Introduction

In recent years, rubber products as engineering materials have been used in many industries such as seals, vehicles tires, hydraulic hoses, and engine mountings. These materials usually exhibit very large strains, with a strongly non-linear stress–strain relation. The wide range of application of hyperelastic materials justifies a high research interest in their characterization through experimental techniques and in the study of accurate constitutive models able to characterize their particular behaviour [1]–[3]. As an important type of polymer material, rubber (also called elastomer) is widely used in industry due to its reversible and high deformability. The introduction of computers in structural analysis using finite element analysis (FEA) allows computer simulation of how a part will respond to loads and forces in its application environment. In other words, it is used to predict the effectiveness of design concepts of elastomeric components. This is mainly because elastomeric components exhibit large deformations, nonlinear stress-strain relationships during load, viscoelastic processes such as heat dissipation and time-dependent behaviour, and nearly incompressible characteristics during deformation [4]–[7]. A stored energy function can define rubber materials as hyperelastic material. The coefficients of these functions should be determined by shear test, biaxial, and uniaxial data.

The biggest problem in characterising the mechanical behaviour of rubber materials is how to determine the proper strain energy function and material model that provides a good fit with available experimental data. In this study, different material models groups, such as Mooney-Rivlin, Principle of stretch and Micromechanical models for incompressible isotropic hyperelastic materials were used to model and capture the hyper-elastic mechanical properties of SBR with different percentage of graphene platelets. The relative percentage error was calculated to determine the goodness of data fit to select the best model that can accurately reproduce the uniaxial tensile test data.

2. Experimental procedures

2.1. Material

The styrene butadiene rubber (SBR, 1502 styrene content 23.5% with a Mooney viscosity of ML (1+4)) at 100 °C =52). The rubber was reinforced with different graphene platelets (GnPs) to prepare samples for tensile tests.

2.2. Experimental work

Compounding GnPs into styrene butadiene rubber was done by a laboratory size open two-roll mill machine at room temperature. The volume ratio of GnPs in composites were 0, 2.4, 5, 10.5 and 16.7 vol.%. The tensile dumb-bell samples of neat SBR and its nanocomposites were die cut to record stress-strain graphs using Instron 5567 tensile machine with 2 kN load cell. The tensile test was accomplished according to ASTM D 412 standard [8]. The sample preparations and experimental procedure details follow a similar process as in Ref. [9].

2.3. Hyperelastic material models for elastomers

The non-linear behaviour of rubber requires specific constitutive material formulations, namely hyperelastic material models. In most commercial finite element (FE) softwares package such as MSC. Nastran, MSC. Marc, Abaqus, Ansys, etc., there are different forms of strain energy density functions W may characterise elastomeric materials. Generally, hyperelastic models are described by strain energy function density (W) which is formulated as a function depending on different magnitudes associated with the strain field and the material constants. Usually:

$$W = W(\lambda_1, \lambda_2, \lambda_3) = W(I_1, I_2, I_3) \tag{1}$$

where λ_1, λ_2, λ_3 are principal stretches, and I_1, I_2, I_3 are the invariants of the green deformation tensor. The strain invariants I_1, I_2 and I_3 in terms of the principal stretch ratios, λ_1, λ_2, and λ_3 are defined as:

$$I_1 = \lambda_1^2 + \lambda_2^2 + \lambda_3^2 \qquad I_2 = \lambda_1^2\lambda_2^2 + \lambda_2^2\lambda_3^2 + \lambda_3^2\lambda_1^2 \qquad I_3 = \lambda_1^2\lambda_2^2\lambda_3^2 \tag{2}$$

where the stretch ratios are defined as:

$$\lambda_i = \left(\frac{L_i + \Delta L_i}{L_i}\right) = 1 + \varepsilon_i \quad ; i = 1, 2, 3 \tag{3}$$

where $\varepsilon_i = \Delta L_i / L_i$ is the engineering strain. In the case of perfectly incompressible materials, the third invariant equals one, i.e. $I_3 = 1$.

Three different hyperelastic models were studied to obtain a constitutive material model for elstomer material, and their formulation are described.

A) Mooney-Rivlin model [10], [11]

$$W = C_{10}(I_1 - 3) + C_{01}(I_2 - 3) \tag{4}$$

where C_{10} and C_{01} are the material constants

IOP Conf. Series: Materials Science and Engineering **244** (2017) 012016 doi:10.1088/1757-899X/244/1/012016

1. Neo-Hookean model [12]

$$W = \frac{1}{2}G(I_1 - 3) = \frac{1}{2}G[(\lambda_1^2 + \lambda_2^2 + \lambda_3^2) - 3] \quad (5)$$

2. Two term Mooney - Rivlin energy function

$$W = C_{10}(I_1 - 3) + C_{01}(I_2 - 3) \quad (6)$$

where C_{10} and C_{01} are the material constants

3. Three term Mooney - Rivlin energy function

$$W = C_{10}(I_1 - 3) + C_{01}(I_2 - 3) + C_{11}(I_1 - 3)(I_2 - 3 \quad (7)$$

where C_{10}, C_{01}, and C_{11} are material constants

4. The Signiorini model

$$W = C_{10}(I_1 - 3) + C_{01}(I_2 - 3) + C_{20}(I_1 - 3)^2 \quad (8)$$

5. Second order invariant

$$W = C_{10}(I_1 - 3) + C_{01}(I_2 - 3) + C_{11}(I_1 - 3)(I_2 - 3) + C_{20}(I_1 - 3)^2 \quad (9)$$

6. Third order deformation (or James-Green-Simpson):

$$W = C_{10}(I_1 - 3) + C_{01}(I_2 - 3) + C_{11}(I_1 - 3)(I_2 - 3) + C_{20}(I_1 - 3)^2 + C_{30}(I_1 - 3)^3 \quad (10)$$

7. Yeoh model [13]

$$W = C_{10}(I_1 - 3) + C_{20}(I_1 - 3)^2 + C_{30}(I_1 - 3)^3 \quad (11)$$

B) Principle of stretch models

1. Ogden model [14]

$$W = \sum_{n=1}^{N} \frac{\mu_n}{\alpha_n}[J^{\frac{-\alpha_n}{3}}(\lambda_1^{\alpha_n} + \lambda_2^{\alpha_n} + \lambda_3^{\alpha_n}) - 3] + 4.5K(J^{\frac{1}{3}} - 1)^2 \quad (12)$$

Usually, the number of terms taken into account in the Ogden models is N = 4. Where μ_n and α_n are material constants, K is the initial bulk modulus, and J is the volumetric ratio defined by $J = \lambda_1\lambda_2\lambda_3$.

2. Foam

$$W = \sum_{n=1}^{N} \frac{\mu_n}{\alpha_n}[(\lambda_1^{\alpha_n} + \lambda_2^{\alpha_n} + \lambda_3^{\alpha_n}) - 3] + \sum_{n=1}^{N} \frac{\mu_n}{\beta_n}[1 - J^{\beta_n}] \quad (13)$$

where μ_n, α_n, and β_n are material constants.

C) Micromechanical models

1. Arruda-Boyce model [15]

$$W = \mu \sum_{i=1}^{N} \frac{C_i}{\lambda_L^{2i-2}}(I_1^i - 3^i) + \frac{1}{D}\left[\frac{J^2-1}{2} - \ln J\right] \quad (14)$$

where C_i and λ_L are material constants, μ is the initial shear modulus and D is a material constant related to the bulk modulus.

$$C_1 = \frac{1}{2}, C_2 = \frac{1}{20}, C_3 = \frac{11}{1050}$$

3. Results & discussion

3.1. Data fitting with different rubber material models

Figure 1. (a) – (c) presented, the result of data fitting with different rubber material models used in the present study. It is clear that the third order deformation model, Figure 1. (b), gave the best curve fitting if it compared with the other models. Moony3 and second order invariant models are shown in Figure 1(a) and (b) gave the second best fit after the third order model. The goodness of fit for the remaining hyperelastic models, Figure 1. (a) –(c) were invariably poor regarding the relative errors of the least square fitting of uniaxial data with these material models, as it will be explained in Figure 2. Therefore, the third order deformation model was chosen to represent the hyperelastic behaviour of the rubber nano-composite materials hereafter in this paper because the relative error does not exceed 3%.

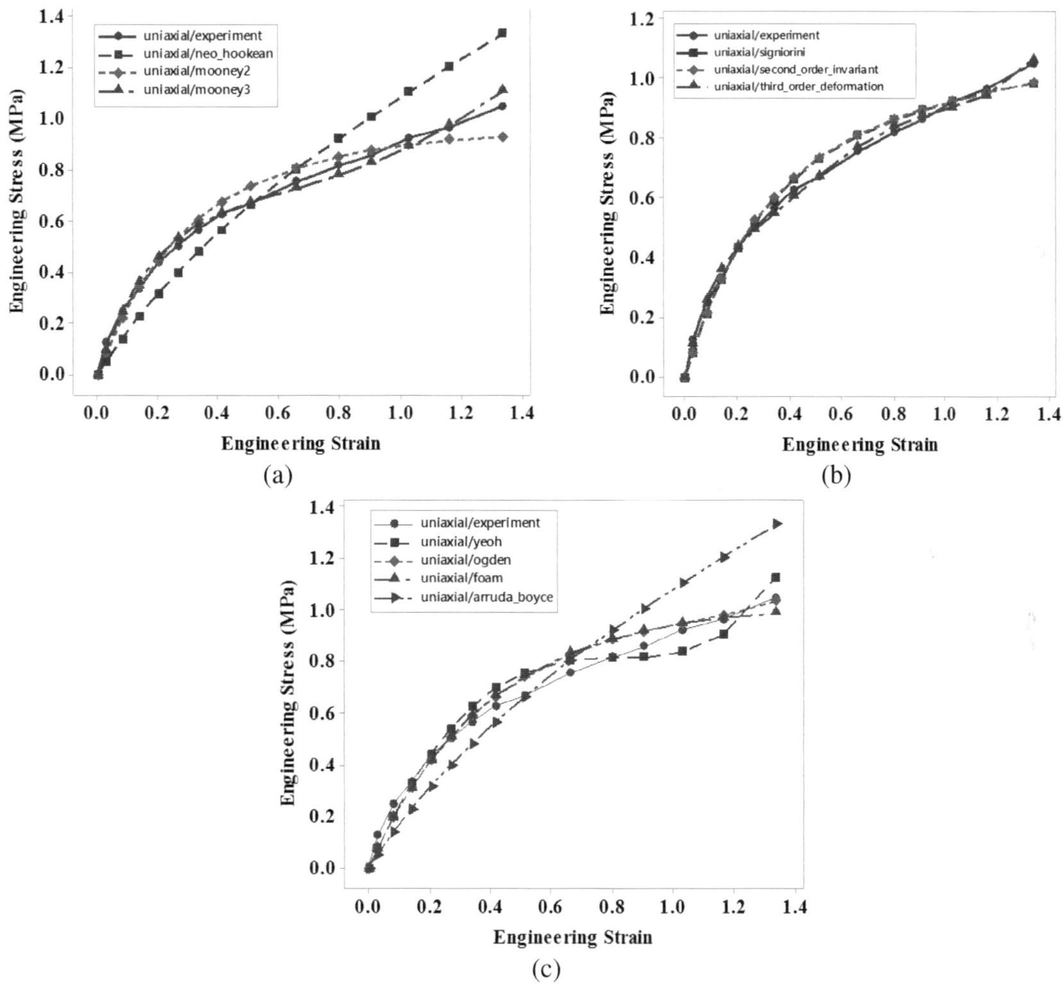

Figure 1. (a)-(c) show the curve fitting of the experimental data with the seven rubber material models, for pure SBR.

3.2. Effect of the ratio of GnPs % on the values of relative error of fitting

Figure 2 shows the variation of percentage relative error values with GnPs% for various material models. As shown in Figure 2, the lowest relative error value was observed with third order deformation model for all GnPs%. Arruda-Boyce model gave the worst fit since the relative error can reach 7%. The value of relative error in all material models increased with the increasing GnPs %. This trend was expected to happen because the stress-strain curves showed a marginal increase with

GnPs%. Figure 3 illustrates the variation of the material constants C10, C01, C11, C20 and C30 of third order deformation model with increasing of GnPs %. The material constants, C20 and C30, slightly affected by GnPs% whereas, C11 and C01 showed a continuous increase with %GnPs especially after 5% GnPs. On the contrary, C10 showed a decreasing trend with the addition of GnPs. The reason for this variation may attribute to the mathematical nature of the third order deformation model which is a polynomial hyperelastic model based on the statistical thermodynamics of cross-linked polymer chains [9-10]. These material constants are empirically determined, and therefore it is difficult to link directly or indirectly their variations to material anisotropy and inhomogeneity of the samples.

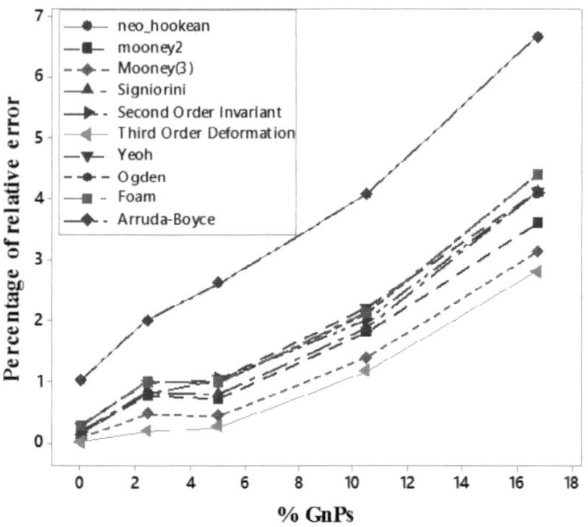

Figure 2. Percentage of relative error of fit for the various material models versus GnPs%.

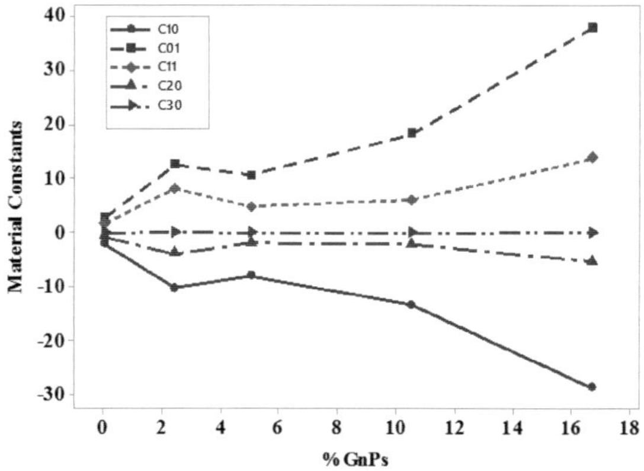

Figure 3. The variation of the material constants for third order deformation model versus GnPs %.

3.3. Comparison between FE-simulation and experimental data

3.3.1. The force/displacement curve.

Figure 4 shows a comparison between experiments and FE prediction for load/displacement response of uniaxial tensile test at different GnPs%. The effect of graphene platelets content, GnPs%, on the response of the uniaxial tensile test was simulated utilising the material constants obtained from the third order deformation model. It is seen that even addition of small values of GnPs% affected the measured stiffness. An increase in the percentage of GnPs% rubber is getting stiffer and the force-displacement response was higher. The differences between measured and predicted force-displacement response became significant at 16.7% GnPs. However, these differences could be eliminated if enough data was available and the material constants were derived from the combined test data. Material constants derived from the synthetic biaxial data provided a better prediction of rubber behaviour in both tension and simple shear modes [16], [17]. From the engineering point of view, differences between experimental and FE results, in Figure 4, even at high GnPs% are acceptable since the maximum relative error did not exceed 7%. The third order deformation model is sufficiently robust enough to regenerate the experimental data for the uniaxial test.

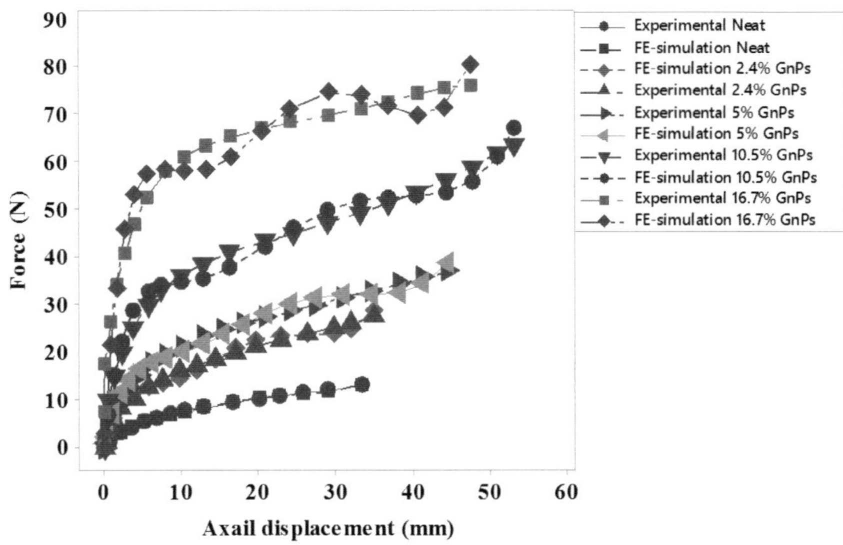

Figure 4. A comparison between experimental and FE- predictions of load/displacement response of uniaxial tensile test at different GnPs%.

3.3.2. Uniaxial tensile strength.

Figure 5 shows a comparison between experimental and FE- predictions of uniaxial tensile strength at various GnPs%. In general, the trend of experimental results and FE- simulations are the same up to 10.5 %GnPs after that a slight difference was observed when GnPs% increased. The reason may be attributed the relative error of fit that increased significantly at GnPs value greater than 10.5 % (see Figure11). For GnPs% values less than 10%, the relative error of fit for third order deformation model was below 1% whereas it reached about 3% at 16.7% GnPs. Regardless of differences between experimental and FE- predictions of tensile strength, still the maximum prediction error did not exceed 4 % even at high GnPs values.

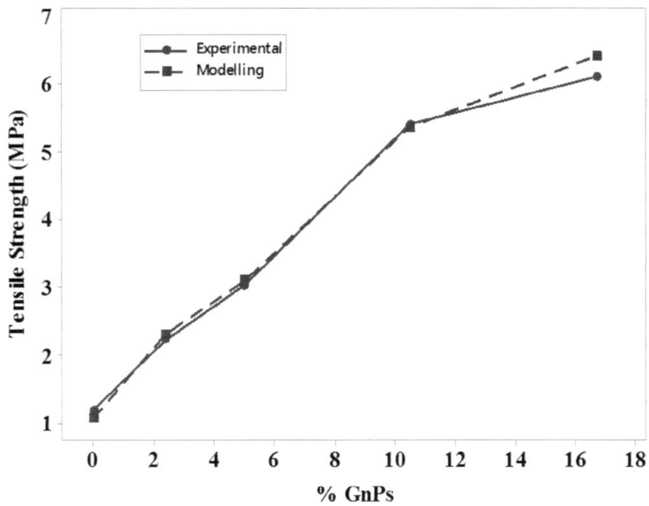

Figure 5. A comparison between experimental and FE- predictions of uniaxial tensile strength at different GnPs%.

4. Conclusion

The aim of this paper was to determine the best material model that representing the tensile behaviour of styrene-butadiene rubber (SBR) at a different volume fraction of graphene platelets (GnPs). To verify that, different hyperelastic material models have been studied in order to obtain the material constants of each model. The relative percentage error has been calculated to judge their goodness of fit and conclude which of them best represents the actual behavior of the rubber nanocomposites. In view of the obtained results, it concluded that the third order deformation model is the best one to regenerate tensile tests of SBR /GnPs samples since the relative errors did not research 2.7 %. Biaxial testing data are needed to minimize the relative error and provide a better prediction of rubber behaviour especially for rubber nano-composites contains GnPs greater than 10.5 vol. %.

5. References

[1] Ali A, Fouladi M H and Sahari B 2010 A review of constitutive models for rubber-like materials *Am. J. Eng. Appl. Sci.* **3** pp.232–239

[2] Sasso M, Palmieri G, Chiappini G and Amodio D 2008 Characterization of hyperelastic rubber-like materials by biaxial and uniaxial stretching tests based on optical methods *Polym. Test.* **27** pp. 995–1004

[3] Hassan M A, Hamdi M and Noma A 2012 The nonlinear elastic and viscoelastic passive properties of left ventricular papillary muscle of a Guinea pig heart *J. Mech. Behav. Biomed. Mater.* **5** pp.99–109

[4] Bhowmick A K 1994 *Rubber Products Manufacturing Technology* 1st ed. (New York CRC Press)

[5] Tobajas R, Ibartz E and Gracia L 2016 A comparative study of hyperelastic constitutive models to characterize the behavior of a polymer used in automotive engines *Proc. Int. Electron. Conf. Mater 2nd.* pp.1-17

[6] Martins P A L S, Jorge R M N and Ferreira A J M 2006 A comparative study of several material models for prediction of hyperelastic properties: application to silicone-rubber and soft tissues *Strain* **42** pp.135–147

[7] Nair A U, Lobo H and Bestelmeyer A M 2009 Characterization of damage in hyperelastic materials using standard test methods and abaqus *2009 SIMULIA Cust. Conf.* pp.1–15

[8] ASTM D 412 – 06a 2007 Standard Test Methods for Vulcanized Rubber and Thermoplastic Elastomers- Tension *Annu. B. ASTM Stand.* pp.1–14

[9] Araby S, Maged A, Zaman I, Ma J and Majewski P 2013 A facile approach to fabricate elastomer/graphene platelets nanocomposites *Int., Conf., on Smart Materials and Nanotechnology in Engineering (SMN) 4th, Gold Coast, Australia.*

[10] Mooney M 1940 A theory of large elastic deformation *J. Appl. Phys.* **11** pp.582–592

[11] Rivlin R S and Saunders D W 1951 Large elastic deformations of isotropic materials. vii. experiments on the deformation of rubber *Philos. Trans. R. Soc. A Math. Phys. Eng. Sci.* **243** pp.251–288

[12] Treloar L R G 1943 The elasticity of a network of long-chain molecules-II *Trans. Faraday Soc.* **39** pp.241–246

[13] Yeoh O H 1990 Characterization of elastic properties of carbon-black-filled rubber vulcanizates Rubber *Chemistry and Technology* **63** pp.792–805

[14] Ogden B Y R W 1972 Large deformation isotropic elasticity - on the correlation of theory and experiment for incompressible rubberlike solids *Proc. R. Soc. A* **326** pp.565–584

[15] Arruda E M and Boyce M C 1993 A three-dimensional constitutive model for the large stretch behavior of rubber elastic materials *J. Mech. Phys. Solids* **41** pp.389–412

[16] Wangkiet R, Sucharitpwatskul S, Sirisinha C, Deeprasertkul C, and Sae-Oui P 2008 Characterization of material constants based on synthetic biaxial data *ScienceAsia*, **34** pp.395–399

[17] Hassan M A, Abouel-kasem A, El-sharief M A and Yusof F 2012 Evaluation of the material constants of nitrile butadiene rubbers (NBRs) with different carbon black loading (CB): FE-simulation and experimental *Polymer* **53** pp.3807–3814

Acknowledgements
The authors would like to thank the Ministry of Higher Education (MOHE) in Egypt for their financial support for achieving the current research. Also, they would like to thank the Japan International Cooperation Agency (JICA) for their continuous support to Egypt-Japan University of Science and Technology (E-JUST).

IOP Conf. Series: Materials Science and Engineering **244** (2017) 012017 doi:10.1088/1757-899X/244/1/012017

An improved edge detection approach and its application in defect detection

Vilas H Gaidhane[1], Navdeep[2], Asha Rani[2] and Vijander Singh[2]

1 Electrical and Electronics Engineering Department, BITS Pilani, Dubai Campus, Dubai International Academic City, UAE
2 ICE Division, Netaji Subhas Institute of Technology, New Delhi, India
E-mail: vhgaidhane@dubai.bits-pilani.ac.in

Abstract. In this paper, a simple and efficient edge detection technique is proposed. It is based on the properties of the hyper smoothing function and the fundamentals of modified local binary pattern and hence, called as HY-LBP. The main advantage of the proposed approach is that the relationship between surrounding pixels and centric one is calculated effectively and extract the surrounding information discriminatively. The counting scheme that counts the number of image points whose pixel values are greater or equal than that of the central pixel help to reduce the noise and blurring effects. Thus, HY-LBP can effectively deal with the noises, blurring, and contrast variation. The effectuality and feasibility of the proposed approach is demonstrated on various synthetic and real time images. Experimental results have shown the effectiveness as well as the better performance of the proposed approach for edge detection as compared to the other existing methods. The presented method can be used in metal sheet defect detection applications.

1. Introduction

Over the last decade, X-rays are broadly used for noninvasive industrial detecting [1-3]. Computed tomography and digital radiography techniques are also popular and used in many applications such as diagnostic of fractures in the human body, defects and cracks in the products [4, 5]. However, such methods are expensive and demands a well trained and expert operator. Edge extraction is a good alternative to such conventional methods and plays an important role in computer vision and image processing applications. It is used in medical image analysis and understanding, remote sensing, object-based image coding, and image retrieval. Traditional edge extraction methods such as Canny and Sobel are amazing for their own special features. However, they are unable to keep the balance between noise elimination and edge preservation or they have to adjust parameters whenever the segmentation results are not satisfactory. Thus, a simple method with nice local property remains the challenge for researchers.

As the demand of edge detection in different applications, it has received considerable attention over the last several decades and numerous novel methods have been proposed in the literature [6-8]. Among these methods, local binary pattern (LBP) approach is one of the simple, less complex, invariant to monotonic gray scale and effective to describe textures of the image. However, LBP is sensitive to noise and it is unable to describe the difference of local region as well as degree of difference. Therefore, many researchers have been extended LBP's binary patterns to trinary patterns (LTP) [9], quinary patterns (LQP) [10] and multiple patterns(LMP) [11], by including the several thresholds into the comparison function. These LBP-based methods are greatly used in real time computer vision application such as emotion recognition and classification, face recognition, fingerprint identification, and automated cell pheno type image classification [12-15]. However, these approaches are based on the histograms with bins which are time intensive and memory consuming processes. In these methods the selection of threshold is crucial as well as difficult task.

Recently, *Bi et al.* [16] proposed an effective edge extraction method using improved local binary pattern (H-LBP) for blurry digital radiography (DR) image. This method is able to describe the relationships between pixels and less dependent on binarization. With a improved counting scheme that counts the number of pixels whose values are greater or equal to the centric one, H-LBP can deal with the noises, blurring, and contrast variation. However, it is observed that the edge extraction results and computational efficiency further can be improved using a new hyper smoothing function. Therefore, an attempt has been made and a simple and effective edge detection method based on hyper smoothing function and improved local binary pattern is proposed.

In this paper, LBP method is improved by embedding a hyper function and a counting scheme, so-called HY-LBP. The proposed approach is able to describe the relationship between surrounding pixels and centric one more accurately. With a counting scheme that counts the number of points whose pixel values are greater than that of the centric one, HY-LBP handle noises, blurring effect and obtained the fine edges.

2. Existing methods

Initially, LBP operator was introduced by Ojala *et al.* [6] for texture analysis and appearance-based face recognition. The LBP operator uses the neighborhoods of $3\times3, 8\times8$ or 16×16 pixel size. In this, operator labels the pixel p_n $(n=0,\ldots,7)$ of an image by thresholding each pixel with the value of the center pixel p_c. The values in each pixel are thresholded with the center pixel value and a binary number is extracted. It is represented as:

$$B(p_i - p_c) = \begin{cases} 1 & \text{if} \quad p_i \geq p_c \\ 0 & \text{otherwise} \end{cases} \tag{1}$$

The LBP can be computed as follows.

$$H_i = \sum_{i=0}^{n-1} B(p_i - p_c)\, 2^i, \qquad \forall\, i = 0,2,3,\cdots,n-1. \tag{2}$$

where 2^i is a binomial factor for each $B(p_i - p_c)$. This method is simple and easy to understand and therefore, it has been extensively used for pattern analysis as well as other applications [7, 10-11]. However, it is observed that the LBP method is more sensitive to noise and it cannot show the degree of difference among the pixel. To overcome these drawbacks, many researches extended the binary patterns to different patters by adding a several threshold into the function [9-11]. Recently, Bi *et al.* [16] carried out an extensive work on edge extraction of blurry digital radiography images and proposed an effective technique based on the H-function commonly called as H-LBP. In this method the number of surrounding pixels are recorded whose gray-scale values are equal or greater than that of the central pixel.

$$\Gamma_i = \sum_{i=1}^{8} B(p_i - p_c),$$

$$B(\eta) = \begin{cases} 1 & \eta \geq 0 \\ 0 & \eta \leq 0 \end{cases} \tag{3}$$

The H- LBP is calculated as [16]

$$H-LBP_{8,1}=\begin{cases}0 & T=0\\ \sum_{i=1}^{8}H_c(p_i-p_c)2^i & 1\le T\le 7\\ 0 & T=8\end{cases} \quad (4)$$

where $H_c(p_i-p_c)$ = compose $H(p_i-p_c)$.

The recorded number T can be used as a threshold to redundant fine gray-scale changes to abandon the isolated points and throw off the points lying in smooth area. The H-LBP approach is demonstrated using Figure 1. It is observed that the H-LBP is more complicated and requires more execution time as compared to existing methods.

91	68	33
33	35	45
20	35	14

↓

0.133	0.099	0
0		0.044
0	0	0

H-LBP = 24

6	5	4
3	5	5
4	7	4

↓

0	0	0
0		0
0	0	0

H-LBP = 0

Figure 1. Example of different regions with unequal H-LBP values.

3. Improved proposed approach

The proposed approach is based on the properties of the hyper smoothing function and the fundamentals of modified local binary pattern. To reduced the sensitivity to noise, a first number of surrounding pixels are recorded whose values are equal to or greater than that of the central pixel.

$$T_h=\sum_{i=0}^{7}S(p_i-p_c),\quad S(\eta)=\begin{cases}1 & \eta\ge 0\\ 0 & \eta<0\end{cases} \quad (5)$$

where T_h is threshold value to redundant fine gray-scale changes. In LBP operation $S(\eta)$ is the comparison function to describe the degree of differences between the pixels. However, this function has some insufficiencies. It is more sensitive to noise and unable to describe the difference between local region. To overcome these problems, a hyper smoothing function is proposed as:

$$HY(\eta)=\frac{1}{2}\left(\frac{1}{2}\log(\chi_i)\right) \quad (6)$$

where $HY(\eta)$is a hyper function in the level set method and acts a step function. The proposed HY-function is calculated as:

$$HY(p_i-p_c)=\begin{cases}0 & p_i-p_c<T_h\\ \frac{1}{2}\left(\frac{1}{\pi}\log(\chi_i)\right) & Otherwise\end{cases} \quad (7)$$

where

$$\chi_i = \left(\frac{p_i + p_c}{p_i - p_c} \right)^{\frac{1}{4}} \tag{8}$$

Thus, HY-LBP with smoother function is calculated as

$$HY - LBP_{8,1} = \begin{cases} 0 & T = 0 \\ \sum_{i=1}^{8} HY_c(p_i - p_c)2^{i-1} & 0 \le \mathrm{T} \le 7 \\ 0 & T = 8 \end{cases} \tag{9}$$

where, $HY_c(p_i - p_c) = compose(HY(p_i - p_c))$ and T is the recorded number aims at throwing off isolated noises and points lying in smooth area. Moreover, $T = 0$ represents the points isolated from noise while $T = 8$ represent isolated points or points lying in smooth area.
The proposed approach is demonstrated as shown in Figure 2. It is observed from Figure 2, that the proposed approach gives the unequal results when the degrees of difference between p_i and p_c are not the same.

91	68	33
33	35	45
20	35	14

↓

0.032	0.045	0
0		0.083
0	0	0

HY-LBP = 8

6	5	4
3	5	5
4	7	4

↓

0	0	0
0		0
0	0	0

HY-LBP = 0

Figure 2. Example of different regions with unequal HY-LBP.

Based on the above mathematical preliminaries a new edge detection method is proposed which is summarized in Algorithm 1.

Algorithm 1 Proposed edge detection approach

Step 1: Given an acquired image $I_p(x, y)$ of size $M \times N$ pixels smoothen by 3×3 mean filter and normalized by its energy defining as:

$$\tilde{I}_p(x, y) = \frac{I_p(x, y)}{\sqrt{\sum_{x,y=0}^{N-1} \left[I_p(x, y) \right]^2}}$$

such that $\sum_{x,y=0}^{N-1} \left[\tilde{I}_p(x, y) \right]^2 = 1.$

Step 2: For each pixel in a 3×3 neighborhood, count the number T_h using Eq. (5).
Step 3: Moreover, in a 3×3 neighborhood, compute the $HY(p_i - p_c)$ function using Eq. (7) for each surround pixel.
Step 4: Short the $HY(\eta)$values by $HY_c(p_i - p_c) = compose(HY(p_i - p_c))$and further weight these values using Eq. (9).

Step 5: Set the current pixel to white if HY-LBP values of the current pixel and previous pixel is non-zero.
Step 6: Repeat Step 2 to Step 5 for the entire image area.

Generally the digital radiography images are full of noises and artifacts which can be eliminated at the initial stage in the proposed algorithm. It can be done using mean, median filters or normalization process on the image. Further the threshold function T_h is used to isolate the residuals of the noises and artifacts from the images.

4. Experimental results and discussion

In this section, the performance of the proposed focus measure is tested by performing a series of experiments on various images. All the experiments are carried out on Intel Core^{TM} i3-370M 2.4, GHZ and 4 GB RAM processor, MATLAB platform of version R 2011.

The proposed approach (HY-LBP) is tested on blurry digital radiography and sheet metal plates images. All the digital radiography images as wll as the sheet metal plates used in the experimentations are full of noises and artifacts. Therefore, the effectuality of the proposed approach is tested without simulating the images at different level of noises.

4.1. Blurred digital radiography (DR) images

In the first experiment, the effectuality of the proposed approach is tested on the blurry digital radiography (DR) images. Recently, Bi *et al.* [16] carried out the experimentations on various DR images. In this experiment same database has been considered for the fair comparison of the results. This database was acquired by scanning the railway side frame of 2286 mm length and 427 mm width and 695 mm height. These images are stored as gray scale images with cropped size of a 5876×912 pixel. An example of such blurry digital radiography images is shown in Figure 3(a).

The comparison of proposed approach HY-LBP, and the various existing edge extractions methods such as H-LBP, Canny, Sobel and LBP alone is shown in Figure 3. To observe the more details in the results, one important parts of casting is considered in the original image shown in Figure 3(a). In this image batch number (B) and low contrast area is seen clearly.

The various experimentations are carried on the original blurry digital radiography image. The obtained results are shown in Figure 3. The proposed approach called as HY-LBP shows the best results as shown in Figure 3(b). It is observed that the HY-LBP clearly shows the batch number (B) and low contrast area. Further, Figure 3(c)-(f) shows the result of H-LBP technique, Canny method, Sobel method, and LBP alone, respectively. It observed from Figure 3(d)-(f) that the batch number, low contrast and edges are not clear. Moreover, the HY-LBP results as shown in Figure 3(b) are comparatively better and clearer as compared to H-LBP technique. Hence the presented approach better than the well known existing methods such as H-LBP, Canny, Sobel and LBP method.

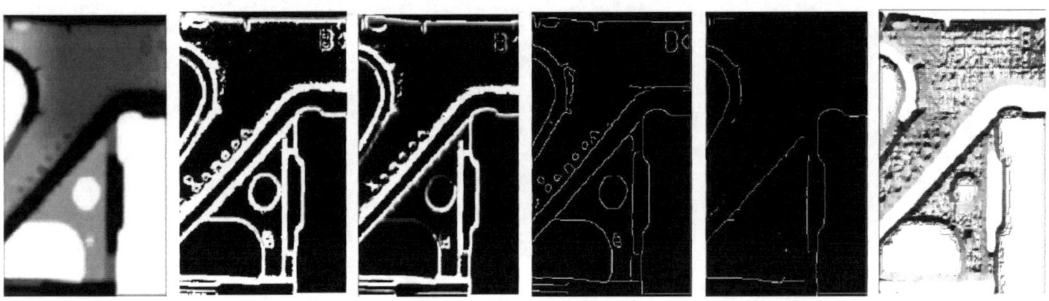

(a) (b) (c) (d) (e) (f)

Figure 3. Comparison of edge extraction by different methods in DR image. (a) Original image (b) HY-LBP (c) H-LBP (d) Canny Method (e) Sobel Method (f) LBP alone.

4.2. Sheet metal plates images
In second experiment, the efficiency of the proposed approach is tested on the sheet metal images. The presented approach can be used to detect the defects and cracks on the metal sheet plates. To demostarte this application, the various experimentations are carried out on the rolling metal sheets images. The rolling defects can be easily visualized and detected using the different edge detection methods. The comparison of presented approach, and the existing edge extraction methods such as H-LBP, Canny, Sobel and LBP alone is shown in Figure 4. It is observed from Figure 4 that the proposed approach HY-LBP results into more clear and accurate cracks and edges which may be useful in identifying the rolling defects on the metal sheets.

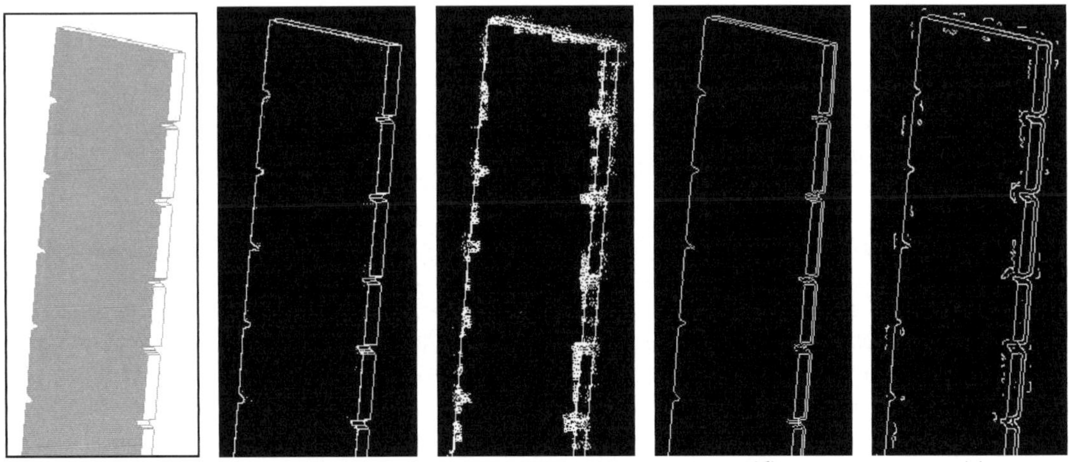
(a) (b) (c) (d) (e)

Figure 4. Comparison of edge extraction by different sheet metal plate image. (a) Original image (b) HY-LBP (c) H-LBP (d) Canny Method (e) Sobel Method.

4.3. Computational time
The comparison of computational time of proposed approach HY-LBP, H-LBP, Canny, Sobel and LBP alone is summarized in Table 1.

Table 1. Computational speed of reported and proposed edge extraction methods.

Methods	HY-LBP	H-LBP	Canny	Sobel	LBP
Time (s)	1.143±0.031	2.741±0.020	0.481±0.016	0.28±0.012	0.134±0.0013

From Table 1, it is observed that the proposed approach requires less time as compared to H-LBP. However, the computational time of the proposed HY-LBP approach is slightly greater than the Canny, Sodel and LBP alone. Although the computational time of proposed approach is more, it is acceptable due to its overall better performance and robustness in the presence of blurred and noisy conditions.

5. Conclusions
In this paper, HY-LBP is proposed as an effective edge detection technique. The performance is improved due to the smoothing hyper function and the similarity distance measure, embedded with the conventional local binary pattern method. With a counting scheme that counts the number of points whose pixel values are greater than that of the centric one, HY-LBP handle the effects such as noise, blurring, and low contrast. Thus, it has high discrimination power in the presence of blurring effect and noises. The experimental results using the synthetic and real time images under blurred and noisy conditions show that the proposed approach performs significantly better in terms of noise robustness, edge discrimination capability, contrast invariance and computational time. Moreover, experimentations on the images show that the HY-LBP performs comparatively better than the existing edge extraction methods.

6. References

[1] Hussein WB, Moaty AA, Hussein MA, Becker T, A novel edge detection method with application to the fat content prediction in marbled meat 2011 *Pattern recognit* **44**(12), 2959–2970

[2] Han XH, Chen YW, A robust method based on ICA and mixture sparsity for edge detection in medical images 2011 *Signal Image Video Process* **5**(1), 39–47

[3] Kersting T, Schonartz N, Oesterlein L, Liessern A, High end inspection by filmless radiography on LSAW large diameter pipes 2010 *NDT & E Int* **43**(3), 206–209

[4] Gaidhane VH, Hote YV, Singh V, An efficient similarity measure approach for PCB surface defect detection 2017 *Pattern Anal Applic* https://doi.org/10.1007/s10044-017-0640-9

[5] Kumar A, Computer-vision-based fabric defect detection: a survey 2008 *IEEE Trans. Ind. Electron* **55**(1), 348–363

[6] Ojala T, Pietikainen M, Harwood D, A comparative study of texture measures with classification based on feature distributions 1996 *Pattern Recognit.* **29**(1), 51–59

[7] Ojala T, Pietikainen M, Maenpaa T, Multiresolution gray-scale and rotation invariant texture classification with local binary patterns 2002 *IEEE Trans. Pattern Anal. Mach. Intell.* **24**(7), 971–987

[8] Ma L, Tan T, Wang Y., Zhang D, Personal identification based on iris texture analysis 2003 *IEEE Trans. Pattern Anal. Mach. Intell.* **25**(12), 1519–1533

[9] Tan XY, Triggs B, Enhanced local texture feature sets for face recognition under difficult lighting conditions 2010 *IEEE Trans. Image Process.* **19**(6) 1635–1650

[10] Nanni L, Lumini A, Brahnam S, Local binary patterns variants as texture descriptors for medical image analysis 2010 *Artif. Intell. Med. Image Anal.* **49**(2), 117–125

[11] Zhu CR, Wang RS, Local multiple patterns based multiresolution gray-scale and rotation invariant texture classification 2012 *Inf. Sci.* **187**(1), 93–108

[12] Gaidhane VH, Hote YV, Singh V, An efficient approach for face recognition 2014 *Pattern Recognit.* **47**(5), 1869–1879

[13] Gaidhane VH, Hote YV, Singh V, Emotion recognition using eigenvalues and Levenberg-Marquardt algorithm-based classifier 2016 *Sadhana: Acad. Proc. Eng. Sci.* **41**(4), 415–423

[14] Kumar A, Chan TST, Robust ear identification using sparse representation of local texture descriptors 2013 *Pattern recognit.* **46**(1), 73–85

[15] Nanni L, Lumini A, A reliable method for cell pheno type image classification 2008 *Artif. Intell. Med.* **43**(2), 87–97

[16] Bi B, Zeng L, Shen K, Jiang H, An effective edge extraction method using improved local binary pattern for blurry digital radiography images 2013 *NDT & E Int.* **53**(1), 26–30

IOP Conf. Series: Materials Science and Engineering **244** (2017) 012018 doi:10.1088/1757-899X/244/1/012018

Thermal deformation analysis of Spliced high energy laser emission window

Xutang Liu [1*], Diansheng Cao[2]and Guangdou Liu [1]

1 College of Mechanical and Electronic Engineering,
China University of Petroleum, Qingdao 266580, China
E-mail: lxt220@upc.edu.cn, 443788006@qq.com,
2 Changchun Institute of Optics and Fine Mechanics and Physics, Chinese Academy of Sciences, Changchun, 130000, China
E-mail: caodiansheng1987@163.com

Abstract. When laser work continuously, laser and external environment will cause damage to the laser window. The paper analysis the working environment of the equipment and the main factors that cause the window damage. Emulate mirror deformation of the laser emission window under the coupling of force and heat, fit the shape change law of different material windows in force thermal coupling by MATLAB. The results show that high energy laser will have a serious influence on the surface microstructure of windows, splicing window can greatly reduce the influence of window deformation on laser emission, ZnSe' characteristic of materials is best suited for producing wide band iraser emission windows.

1. Introduction

High energy laser is widely used in laser drilling, Machining and Material surface modification. window is the key component to ensure the normal operation of laser emission equipment[1,2]. The window protects the interior of the device from the outside environment[2-4],enhances the environmental adaptability of the equipment and improves the service life and reliability of the equipment. The thermal deformation occurs when the window absorbs the laser energy, which causes the wavefront distortion [4,5], the center shift of the laser beam and affects the laser beam quality and transmission characteristics. When the thermal damage beyond the material damage threshold will produce a series of irreversible, such as oxidation, hot melt breakage[6-8], stressand microstructure of the material will produce corresponding changes such as the change of phase inter phase coupling and micro crack propagation[9,10]. Splicing window can effectively improve the stiffness of the window, reduce the window thickness, increase the window cooling and reduce the laser heat deposition and window temperature rise. Paper analysis a splicing high energy laser emission window according to actual engineering needs, the temperature distribution of splicing window and ordinary window is simulated by finite element software, The material property of common light materials in windows are analysed, the thermal deformation of different material windows is fitted by MATLAB, The results of the correlation analysis can provide the design basis for the engineering application of the splicing window.

2. Window design and calculation

2.1. Window structure design

The basic structure of the splicing window is shown in Figure 1, the window is made up of 5 SnSe lenses, the center is a circular lens with a diameter of 160mm, The outer layer is four inner diameter 80mm outer diameter 320mm fan-shaped lens, The round lens has low laser capacity and is rigidly fixed. The fan-shaped lenses bear high laser energy and are supported by flexible hinges to reduce the thermal stress of the lenses.

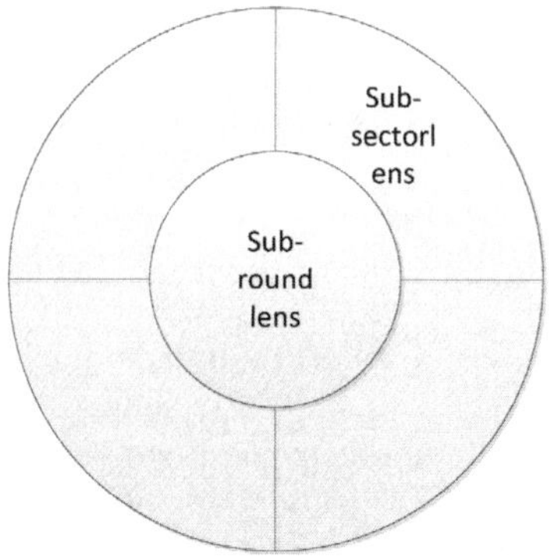

Figure 1. Sketch map of spliced laser emission window.

2.2. Analysis of laser emission windows working environment

The working environment of the splicing window is shown in figure 2, After the high energy laser beam enters the laser launcher, 80% of the capacity is reflected into the sector lens through the primary mirror, 20% of the energy passes through the center of the mirror into the round lens. The working environment of the high energy laser emission system is shown in Figure 3, the inner window is mainly affected by high energy laser beam, the outside is mainly affected by the environmental medium in the equipment working environment, When the window surface temperature is lower than the material damage threshold, the influence of high energy laser beams on windows is dominated by thermal deformation, When the temperature in the irradiated area reaches a certain critical point, irreversible oxidation such as oxidation, melting and gasification will occur in the window. The mirror damage on the inner side of the window is mainly caused by the thermal deformation of the high energy laser beam, Outside the window is the atmosphere, the air flow quickly and equipped with cooling air to assist cooling, the mirror surface thermal deformation can be neglected.

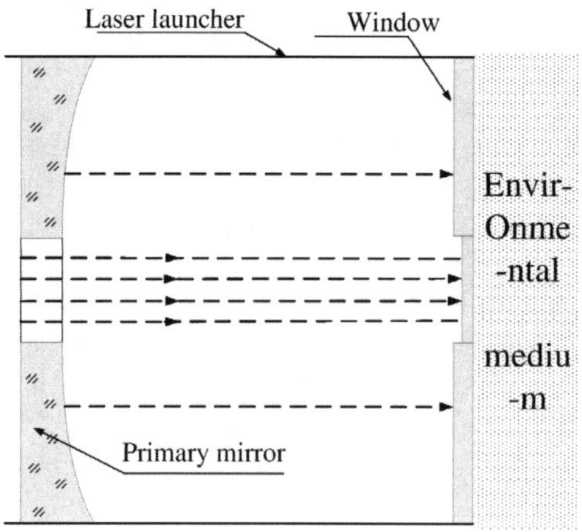

Figure 2. Spliced window working environment.

2.3. Analysis of influence of machining accuracy on laser beam

Mosaic window is spliced by lens, when there is a parallel deviation behind the window, it will produce a prism effect on the emitted laser, which will affect the pointing accuracy of the laser, The relation between the divergence angle θ of the beam and the window thickness and the parallelism between the front and back surfaces is shown in Figure 3, α is the angle of inclination resulting from the installation error of the window, $\alpha+\beta$ is The deviation of the parallel between the front and back surfaces of the window, r_i and r_3 is Incidence angle, r_2 and r_o exit angle, r_4 is divergence angle, h is he thickness of the window at the incident point.

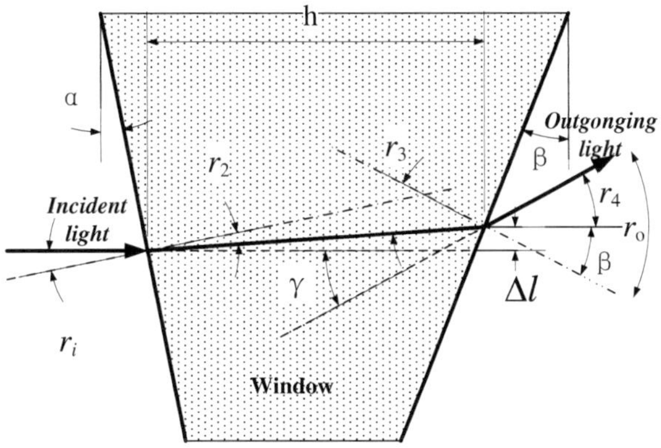

Figure 3. Affect the surface shape of the beam window

According to the sine theorem:

$$\frac{\sin r_2}{\sin r_i}=\frac{\sin r_3}{\sin(r_o)}=\frac{n_0}{n_1} \quad (1)$$

The actual values of each angle in Fig. 3 are small, therefore equation 2 can be simplified as:

$$\frac{r_2}{r_i}=\frac{r_3}{r_o}=\frac{n_0}{n_1} \quad (2)$$

According to the relationship between the outside and inside of the triangle:

$$r_3=\alpha+\beta-r_2=\alpha+\beta-\frac{n_0}{n_1}r_i \quad (3)$$

$$r_0=\frac{n_1}{n_0}\beta+(\frac{n_1}{n_0}-1)\alpha \quad (4)$$

$$r_4=\frac{(n_1-n_0)(\beta+\alpha)}{n_0} \quad (5)$$

In equation n_1 is refractive index of laser beam in window material, ZnSe's refractive index of the output laser beam is about 2.328, n_0 is refractive index of laser beams in air, It can be seen by type 5: the beam divergence angle is about 1.328 times of the parallelism error before the window, which is independent of the window thickness and the installation error.

3. Thermal deformation simulation of laser emission window

3.1. Boundary condition analysis

Window surface heat flux:

$$\begin{aligned} q &= h_{gs}(T_c-T_0)+\alpha\varepsilon\sigma(T_c^4-T_0^4) \\ &= \frac{k}{A}\alpha\left(\frac{g\beta cT_s-T_0}{v}\frac{\rho c_k}{k}\right)^{1/4}\cdot\left(T_c-T_0\right)+\alpha\varepsilon\sigma(T_c^4-T_0^4) \end{aligned} \quad (6)$$

In equation ρ is air density, c is specific heat of air, T_0 is initial temperature of environment, T_c is actual temperature of window surface, α is radiation coefficient, ε is emissivity, σ is *stefan-biltamann* constant， σ=5.669×10^{-4}W/(cm^{2}K^4)， h_{sg} is window surface heat exchange coefficient. According to equation 6, the main factors affecting the heat flux on the window surface are the thermodynamic properties of the material, the velocity of the air at the window surface and the power density of the emitted laser , There are 5m/s cooling air flow both inside and 15m/s cooling air flow outside the window, Therefore, when the laser power density is fixed, the main factor affecting the heat flux of the window surface is the thermodynamic properties of the window material.

3.2. Light intensity distribution of high energy laser beam

The laser for high order Gauss beams, the distribution of electric field strength meet the electric field distribution of Hermite Gauss beam：

$$E(x,y,z)=\frac{A_{mn}w_0}{w(z)}H_m\sqrt{2}\frac{x}{w(z)}H_n\sqrt{2}\frac{y}{w(z)}$$
$$=\exp\left[-\frac{x^2+y^2}{w^2(z)}-i\frac{k(x^2+y^2)}{2R(z)}\right]\cdot\exp\{-i[kz-\phi(z)]\} \quad (7)$$

In equation w is emission laser spot radius，The distribution law of laser light intensity and electric field intensity satisfies the following relation：

$$I_0=\frac{1}{2}\left(\frac{\varepsilon_0 n^2}{\mu_0}\right)^{1/2}E^2 \quad (8)$$

In equation， n is refractive index of a window emitting laser， ε_0=8.854×10-12F/M， is Positive dielectric constant of a laser beam； μ_0 is vacuum permeability, Refractive index, vacuum dielectric constant and vacuum permeability are constant, Therefore, the intensity distribution of the window surface is proportional to the square of the electric field intensity of the emitted laser. The power density distribution of laser can be obtained by MATLAB fitting, as shown in Figure 4.

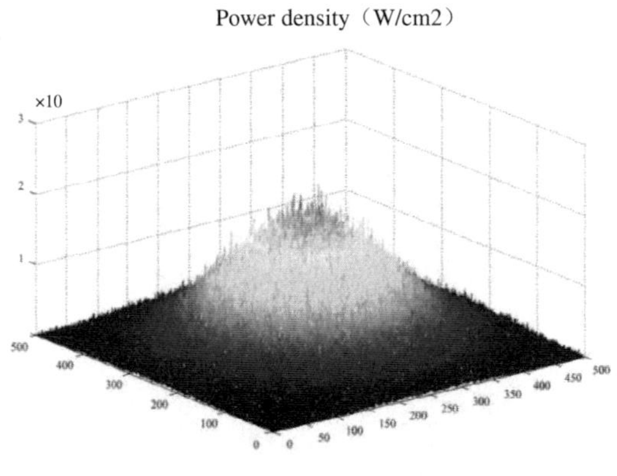

Figure 4. Emission laser power density distribution

3.3. Thermal mechanical coupling analysis and surface shape matching

First of all, assume the ambient temperature is 55 ℃, the window material is ZnSe, the absorption coefficient of the light source is 5 x 10^{-5}/cm, the absorption coefficient of the surface coating is 2.3 x 10-3, the laser power is 11000 watts, the 10.6 m laser is 10000 watts and the 6.6 m laser is 1000 watts, According to the distribution law of laser power density calculated by equation 8，The thermal load is applied to the window physical model to fit the temperature field of the thermal deformation of the window, Heat cell types are converted into structural units using sequential coupling method, the gravity load G, the wind pressure load Q and the corresponding boundary condition are applied to the finite element model, Finally, the distribution pattern of the splicing window and the common window surface is fitted, as shown in Figure 5

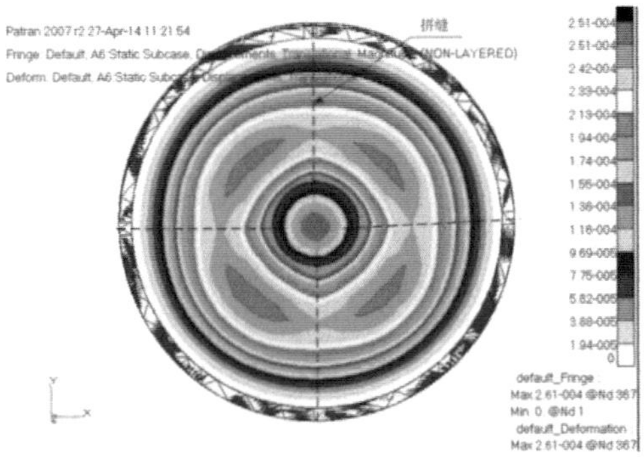

Figure 5. Thermal deformation distribution of spliced window

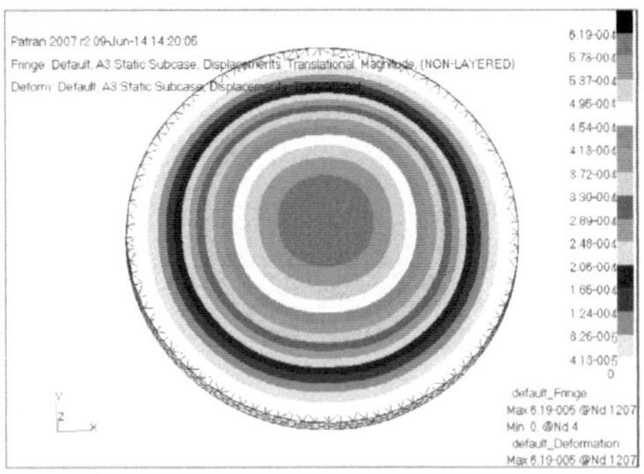

Figure 6. General window thermal deformation distribution

As can be seen from Fig5: The center deformation of sector lens is larger and the deformation of edge is smaller, The main reason is that the intensity distribution of laser is too strong and weak. The deformation of each sub lens' edges is small, and the deformation of the fan-shaped lens' edge is approximately zero, mainly because of the mechanical stress of the clamping mechanism. Due to the occlusion of the laser beam by the secondary mirror bracket, the shape of the window surface is not uniformly distributed along the corner direction, The amplitude of the window shape along the corner is less than 0.1μm. Using the same method, the surface distribution of the conventional whole window is obtained, as shown in Figure 6. As shown in Fig6: the change of the shape of the conventional whole window is about two times that of the mosaic window, The maximum deformation is at the center of the window, this is mainly because the window center is slow to cooling and the heat is easy to assemble.

4. Thermal deformation analysis of infrared window materials

Test results show: the properties of common window materials and materials used in wide band infrared laser transmitting windows are shown in Table 1,

Table1. Materials for Infrared Windows: Properties and Performance

Material	Density (g / cm^{-3})	Melting point (℃)	Coefficient of thermal expansion (1/k)	Thermal conductivity (W/m·k^{-1})	Wavelength range (μm)	absorption coefficient (/cm)	Thermal coefficient (dn/dt)
ZnSe	5.27	1525	7.1e^{-6}	18.0	0.5～22	2e^{-4}	6.1e^{-5}
ZnS	4.09	1827	6.6e^{-6}	16.7	1～14	3.2e^{-4}	4.1e^{-5}
CaF_2	3.18	1360	18.8e^{-6}	9.71	0.13～10	2.2e^{-4}	15e^{-6}
GaAs	5.32	1238	5.7e^{-6}	55	1～14	1e^{-2}	13e^{-5}
Ge	5.33	936	6.1e^{-6}	58.61	2～12	2.3e^{-2}	4.1e^{-4}

Table 1 shows in many window materials: ZnSe absorption coefficient is the lowest, through a wide range of wavelength, the best overall performance. ZnS have the highest melting point, coefficient of thermal expansion is smaller than ZnSe, but absorption coefficient is bigger than ZnSe. GaF_2 have the lowest transmittance after ZnSe in near infrared band, but GaF2's Coefficient of thermal expansion is high. GaAs and Ge's optical properties are lower than that of the above three kinds of materials. The thermal deformation of the window is analyzed by using the characteristics of the materials mentioned above, and the deformation is fitted by the Zernike polynomial. The distribu-tion law of the shape of the window surface along the radius is fitted as shown in Figure 7.

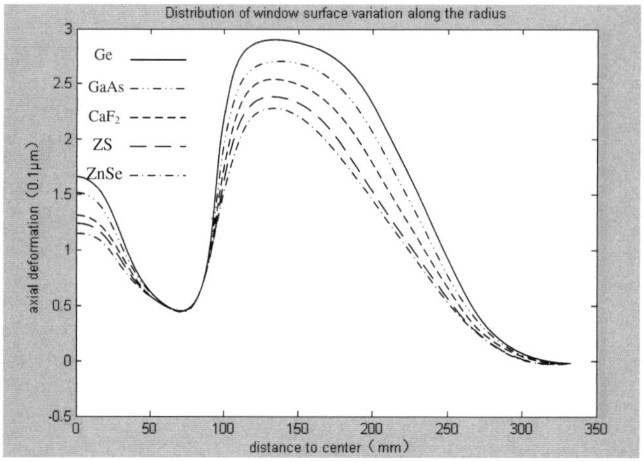

Figure 7. Thermal deformation of different material splicing window.

5. Conclusions

According to the above analysis, we can draw the following conclusions: 1) The thermal deformation of the splicing window is only half of the normal window, 2) The laser emission window made by ZnSe has the least thermal deformation, followed by ZnS, and and CaF2, and GaAs, and and Ge have the most serious thermal deformation.

6. References

[1] Khann, Abasn, Kalaira 2015 Lasers in Engineering,**22** 11

[2] Fei Yan, Yafei Gu, Yajun Wang 2013 Optics&Laser Technolgy**54** 303

[3] Zhichao Liu; Yi Zheng; Feng Pan 2016 Applied Surface Science **69** 294

[4] Z Wei, W Zhang, D Wang, GY Jin 2017 Optik - International Journal for Light and Electron Optics **131** 110

[5] Yaguo Li; Zhigang Yuan; Jian Wang 2016 Optics and Laser Technology **91** 149

[6] N. Lameche; S. Bouzid; M. Hemici 2016 Optik - International Journal for Light and Electron Optics **127** 9663

[7] Luo Wei, Dong Wenfeng, Yang Huabing 2013 Laser & Infrared **43** 845 .

[8] Zhu Xiaohua, Yi Jing, Liu Qingyou2015 science direct **27** 884 .

[9] Peng Hanxiu, Gu Yafei, Wang Chunming 2013 Applied Laser **33** 525

[10] R.H. Zhu,C.M. Li,L.X. Chen,J.L. Liu, J.J. Wei,L.F. Hei,H. Guo 2015 Applied Surface Science. **355** 203

Acknowledgments

Fundamental Research Funds for the Central Universities (15CX02123A). The authors greatly appreciate the financial support of the National Natural Science Foundation of China (51405511) and the Fundamental Research Funds for the Central Universities (16CX02005A)

IOP Conf. Series: Materials Science and Engineering **244** (2017) 012019 doi:10.1088/1757-899X/244/1/012019

Motion performance analysis and control mode design for the cross slide

Dong Xie[1], Jianqu Zhu[1], Min Wang[2], Zhengzhong Li[1] and Feng Wang[3]

1 School of Electrical and Information Engineering, Chongqing University of Science and Technology, Chongqing, China
2 Institute of Continuing Education, Chongqing University of Science and Technology, Chongqing, China
3 College of Mechanical and Power Engineering, Chongqing University of Science and Technology, Chongqing, China
E-mail: xiedongcq@126.com

Abstract. The motion control system which is constructed by using the programmable controller and the motion controller has been widely applied to various mechanical systems. This paper takes a cross slide as an example, the motion control system is composed of the GE FANUC PAC and DSM324i, and the movement performance simulation model is established. The motion performance by linear acceleration and deceleration algorithm in the case of two axis synchronous movement is analyzed. The relation between the displacement command and the parameters of the acceleration and the velocity is derivated, and the trajectory curve of movement performance under a variety of instruction is gained. The control instruction meet the linear motion is designed by the model.

1. Introduction

With the new motion controller appearance constantly, the motion control system composed of the motion controller and the programmable controller (PLC) which can meet customer demand has been widely used in mechanical system, such as the fixed-length cutting machinery, the leather cutting machinery, the carton customized mechanical and the rotating knife cutting machine, etc. This kind of motion control system mainly used for the positioning control of variable distance [1], in addition to achieve a single axis of motion control, also can realize multi-axis linkage control to complete some contour control requirements.

The cross sliding table is consisted of two groups of slider, ball screw and rail. One shaft is mounted on the other axis to form a cross. It can be moving along X, Y direction respectively, or completing the corresponding curve trajectory by two axis synchronous motion. The motion control system using DSM324i and PAC System RX3i programmable controller, can solve the complex single axis and multi-axis motion control [2]. The the block diagram of system structure is as Figure 1.

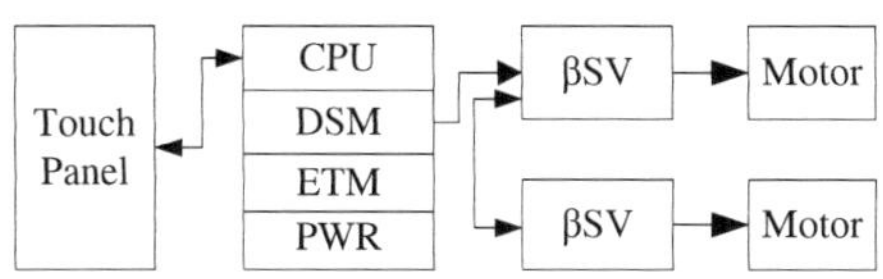

Figure 1. Block diagram of system structure.

IOP Conf. Series: Materials Science and Engineering **244** (2017) 012019 doi:10.1088/1757-899X/244/1/012019

First, the control instruction is programmed on the internal of PAC. And then PAC instructions will be sent to motion controller. The motion position and speed instructions will be converted to servo control signal to drive the servo motor moving.

The motion program is created through programming design of DSM324i. Each motion program is required to set the axis acceleration, the maximum speed, the given target point and acceleration model. When the corresponding exercise program is triggered, the motor will be started. The two axis movement program is as follows.

```
PROGRAM 1 MULTI-AXIS
   ACCEL AXIS1 (Acc1)
   ACCEL AXIS2 (Acc2)
   VELOC AXIS1 (Vmax)
   VELOC AXIS2 (Vmax)
SYNC
   PMOVE AXIS1 (Xp), ABS, LINEAR
   PMOVE AXIS2 (Yp), ABS, LINEAR
ENDPORG
```

Where, Acc1 and Acc2 are the values of the acceleration, the Vmax is the maximum feed rate of the shaft, the Xp and Yp is the target point of the location. The synchronous movement instructions are used here.

In order to investigate the cross slide movement performance, a simple path are designed as shown in Figure 2 in the dotted line. Where the path 1 is the uniaxial motion. The path 2 is the two axes linkage movement along the 45° oblique line. The path 3 and 4 are the two axis movement in different quadrants.

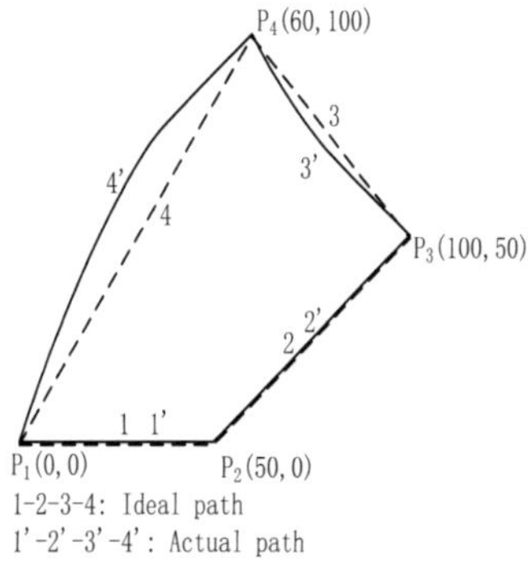

Figure 2. The trajectory of two axis synchronous mode.

The synchronous movement is a simple synchronous movement under the uniaxial motion control. Duo to the axis movement is subject to their maximum velocity and acceleration respectively, its trajectory is irregular, which appeared in Figure 2 lines 1'-2'-3'-4'.

It is seen that the trajectory of uniaxial and 45° oblique line motion can meet the requirements of motion control, and the other two kinds of trajectory appeared larger deviation compared to the theory trajectory.

2. The movement performance simulation mathematical modeling

Because the shaft control is respective independent in the motion controller, its motion is in accordance with the time of itself acceleration [3-5]. Synchronous instruction is only synchronized with the start of the movement, and is meet the requirements of the deceleration to reach the target at the same time, which leads to the movement of each axis will be calculated in accordance with its own parameters. Therefore, according to the different displacement of the shaft, three kinds of situations are discussed. Where, the axis which the target point project on the coordinate has longer motion displacement is called the long axis, the other is the short axis.

IOP Conf. Series: Materials Science and Engineering **244** (2017) 012019 doi:10.1088/1757-899X/244/1/012019

The displacement of the two axes is long enough. Both the long axis and the short axis can complete a whole process including acceleration and deceleration.
Only one of the axial displacement is sufficient. That is, the long axis can be completed a complete acceleration and deceleration process, and the short axis can not reach its maximum speed value.
The displacement of the two axes is not long enough. Both the long axis and the short axis can not reach its maximum speed value.

2.1. Both two axes can run a whole acceleration and deceleration process

The independent control shaft synchronous movement must be abide by the two rules.
The first rule is the priority of slowing down, that is, no matter how much the long axis displacement difference from the short axis's, they must be stopping at the same time (the velocity must be slowing down to zero at the same time).
The second rule is started at the same time, the acceleration is preferred. When a given acceleration, speed instruction cannot be achieved at the same time, give priority to meet the same acceleration, and the maximum velocity can be different. The velocity curves of the long axis and short axis are shown as Figure 3. V_L is long axis velocity, V_S is short axis velocity.

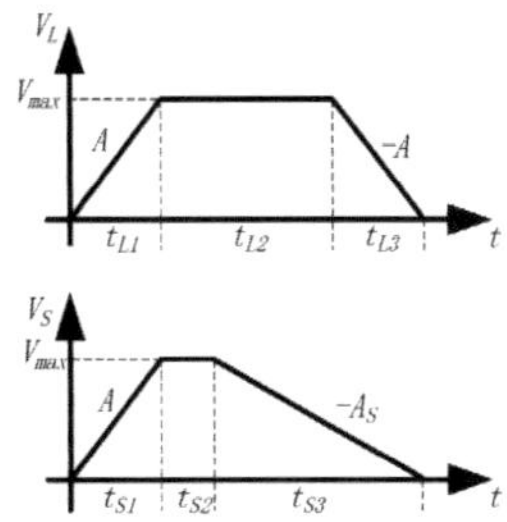

Figure 3. The velocity curve schematic.

Where t_{L1}, t_{L2}, t_{L3} is respectively the three periods of the acceleration time, constant speed time, deceleration time of the long-axis . The t_{S1}, t_{S2}, t_{S3} represents three stages of the short-axis. The trapezoidal area is corresponding to the displacement of major axes S_{Long} and minor axes S_{Short}.
Due to the acceleration and deceleration times are the same in long-axis, so $t_{L1} = t_{L3} = V_{max}/A$. Long axis speed curve is a equilateral trapezoid, the trapezoidal area of long axis is the total displacement as follow:

$$S_{Long} = (t_{L2} + t_{L1})V_{max} \tag{1}$$

The total times t is:

$$t = \frac{S_{Long}}{V_{max}} + \frac{V_{max}}{A} \tag{2}$$

So $t_{L2} = t - 2t_{L1}$. The short axis total movement time is the same as the long axis'. which have complete three sections of curves, so the first stage of the short axis is the same as the the long axis. The minor axis $t_{S1} = t_{L1}$. The short shaft speed curve area is its displacement, so t_{S2}, t_{S3} can be solving as following function group.

$$\begin{cases} S_{Short} = \frac{1}{2}t_{S1}V_{max} + t_{S2}V_{max} + \frac{1}{2}t_{S3}V_{max} \\ t = t_{S1} - t_{S2} - t_{S3} \end{cases} \tag{3}$$

According to t_{S3} can calculate the minor axis deceleration value.

$$A_S = \frac{V_{max}}{t_{S3}} \tag{4}$$

2.2. Only one of the axial displacement is sufficient

Only one axis is long enough, the long axis can complete three stages moving, but the short axis only exist two stages, that is, acceleration and deceleration phase, there is no uniform phase. According to the deceleration priority principle, it can be known that the long axis and the short axis must simultaneously decelerate to zero, so the total time is the same. The short axis of the maximum speed is unknown. Set the maximum speed of the short axis is V_{smax}.

$$v_{smax} = \frac{2S_{Short}}{t} \tag{5}$$

According to the principle of starting priority, the starting stage acceleration is still the acceleration A, then:

$$t_{s1} = \frac{v_{smax}}{A} \tag{6}$$

The short axis deceleration time is $t_{S3} = t - 2t_{S1}$, $t_{S2} = 0$, and the short axis deceleration:

$$A_s = \frac{v_{smax}}{t_{s3}} \tag{7}$$

2.3. Neither of axes can run a whole acceleration and deceleration process

As the long axis and short axis motion path is short, there is no uniform motion phase but the acceleration and deceleration process. It can't accelerate to the maximum speed. Set the long axis has the same period of acceleration and deceleration, $t_{L1} = t_{L3}$, $t_{L2} = 0$.

$$t_{L1} = t_{L2} = \frac{1}{2}\sqrt{\frac{4S_{Long}}{A}} \tag{8}$$

$$v_{Lmax} = \frac{A}{2}\sqrt{\frac{4S_{Long}}{A}} \tag{9}$$

Similarly, according to the principle of deceleration priority, the short axis movement parameters are calculated by the formula above.

Set Figure 2 curve motion parameters: the maximum velocity is 50mm/s, the acceleration is 100mm/s2, the acceleration model is linear acceleration, its speed curve simulation is shown in Figure 4.

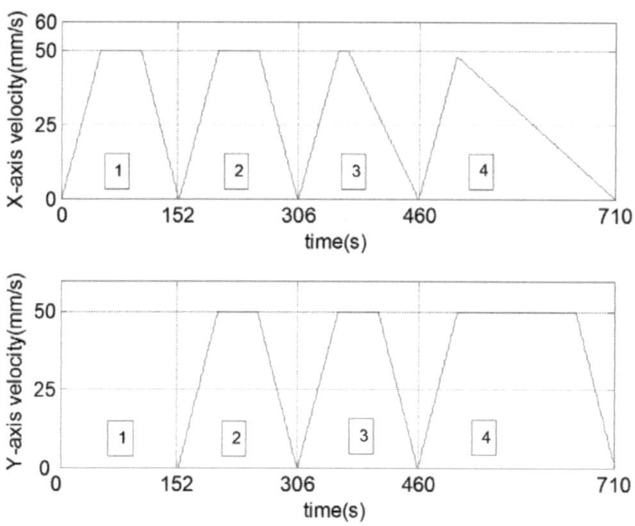

Figure 4. The velocity curve of the trajectory.

3. Improved design of control instructions

In order to realize the linear trajectory motion, the motion parameters of the short axis need to be modified according to the motion state of the long axis. So the linear slope of the straight line is kept constant during the synchronous movement. That is,

$$K = \frac{S_{Long}}{S_{Short}} = Const \tag{10}$$

Where, taking the first case as an example, in the three time periods, the short axis acceleration and maximum speed will have a corresponding decline in order to meet its displacement distribution ratio, then:

$$\begin{cases} S_{Long} = \frac{1}{2} A_L t_{L1}^2 \\ S_{Short} = \frac{1}{2} A_S t_{S1}^2 \end{cases} \tag{11}$$

Where, $t_{L1} = t_{S1}$, $A_L = A$, $S_{Long}/S_{Short} = K$, so,

$$A_S = \frac{S_{Short}}{S_{Long}} A = \frac{A}{K} \tag{12}$$

For the constant speed section,

$$\begin{cases} S_{Long} = V_{L\max} t_{L2} \\ S_{Short} = V_{S\max} t_{S2} \end{cases} \tag{13}$$

Where, $t_{L2} = t_{S2}$, $V_{Lmax} = V_{max}$, $S_{Long}/S_{Short} = K$.

$$V_{S\max} = \frac{S_{Short}}{S_{Long}} V_{L\max} = \frac{V_{\max}}{K} \tag{14}$$

The deceleration phase calculation is the same as the acceleration phase calculation.
The motion control parameters of acceleration and the maximum speed of the short axis can be designed according to the ratio of slope value.The velocity curve of the improved design of control instructions is shown in Figure 5. It is seen that the section 3 and 4 in X-axis is different from that is in the Figure 4. It results a symmetric velocity curve in X-axis.

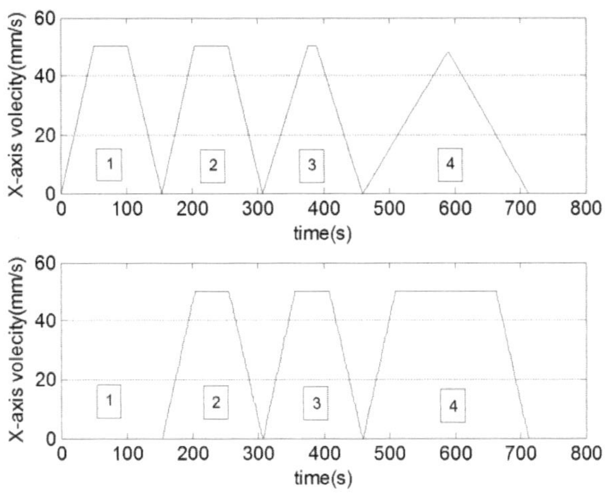

Figure 5. The velocity curve after improved design of control instructions.

IOP Conf. Series: Materials Science and Engineering **244** (2017) 012019 doi:10.1088/1757-899X/244/1/012019

Actual motion path is also consistent with the instruction value, after doing the improvement on the design of control instruction. It is shown Figure 6. It is found that the 3 and 4 segment is not existed the path of the large deviation which in Figure 2.

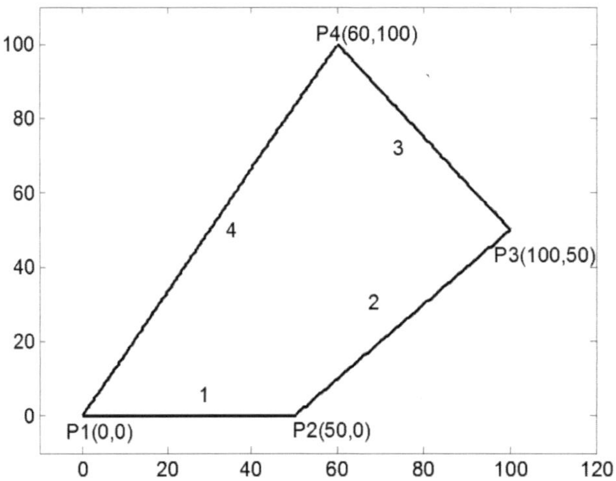

Figure 6. The motion path after improved design of control instructions.

4. Conclusions

In some motion control systems, because of the absence of interpolation, the influence of motion parameters of each axis and motion coordination must be taken into account in the control designing for multi-axis motion. Otherwise it will lead to the deviation of the moving trajectory and acceleration and deceleration incongruity, and resulting in not running in accordance with the ideal state of motion.

5. References

[1] Ye Jun, Yu Xia.Motion Control System Based on GE Fanuc PLC and DSM324i. Coal Mine Machinery. 2011. Vol.32, No.08: 235-236

[2] GE Fanuc Automation. DSM324i Motion Controller for PACSystems™ RX3i and Series 90™-30 User's Manual, GFK-2347A. August 2005

[3] Wu Zhipeng, Chen Xinglin. Design of coarse-fine combined synchronization control system for lithography based on cross-coupled sliding mode control. Advanced Materials Research, Manufacturing Science and Technology, 2012. Vol.383-390:2126-2131

[4] Shi C, Zhao T, Ye P, et al. Study on S-shape curve acceleration and deceleration control on NC system [J]. China Mechanical Engineering, 2007, Vol.18,No.12: 1421-1425

[5] Zhao Ya-Nan.The Study of the Acceleration/Decelaration Mode of Single-axe Motion. Development & Innovation of Machinery & Electrical Products. 2011.Vol.24,No.5:176-178

[6] Chou Po-Huan, Lin Faa-Jeng, Chen Chin-Sheng, et al. DSP-based cross-coupled synchronous control for dual linear motors via intelligent complementary sliding mode control.2012 IEEE Colloquium on Humanities, Science and Engineering Research, 2012, pp266-271

[7] Chi Dongnan.Cross-coupled synchronization control for dual driven scanning system sliding mode-based. Proceedings of the 2015 27th Chinese Control and Decision Conference, July 17, 2015.pp760-765

[8] Sun Mengfei, Li Jianxiong, Fang Yiming.Adaptive cross-coupled sliding mode control for synchronization of dual-cylinder electro-hydraulic servo systems. Proceedings of the 34th Chinese Control Conference, CCC 2015,September 11, 2015,p3361-3366

[9] Liu Chuhui, Huang Fengli.Method of Acceleration/Deceleration Control for High Speed and High Accuracy Machining.Coal Mine Machinery.Apr.2006,Vol 27,No.4:641-643

[10] Li, Hongke, et al. "Design of global sliding-mode controlled AC servo controller based on exponential acceleration/deceleration algorithm." International Conference on Mechatronics and Automation IEEE, 2010:1507-1511

Acknowledgment
This research is funded by Research Foundation of Chongqing University of Science & Technology, the project No. is CK2015Z30.

IOP Conf. Series: Materials Science and Engineering **244** (2017) 012020 doi:10.1088/1757-899X/244/1/012020

Social aspects of automation: Some critical insights

Ibrahim Nouzil [1], Ali Raza [2] and Salman Pervaiz [1]

1Department of Mechanical and Industrial Engineering, Rochester Institute of Technology – Dubai, P.O. Box 341055 Dubai, United Arab Emirates
2Department of Electrical Engineering and Computing Sciences, Rochester Institute of Technology – Dubai, P.O. Box 341055 Dubai, United Arab Emirates
Email: sxpcad@rit.edu

Abstract. Sustainable development has been recognized globally as one of the major driving forces towards the current technological innovations. To achieve sustainable development and attain its associated goals, it is very important to properly address its concerns in different aspects of technological innovations. Several industrial sectors have enjoyed productivity and economic gains due to advent of automation technology. It is important to characterize sustainability for the automation technology. Sustainability is key factor that will determine the future of our neighbours in time and it must be tightly wrapped around the double-edged sword of technology. In this study, different impacts of automation have been addressed using the '*Circles of Sustainability*' approach as a framework, covering economic, political, cultural and ecological aspects and their implications. A systematic literature review of automation technology from its inception is outlined and plotted against its many outcomes covering a broad spectrum. The study is more focused towards the social aspects of the automation technology. The study also reviews literature to analyse the employment deficiency as one end of the social impact spectrum. On the other end of the spectrum, benefits to society through technological advancements, such as the Internet of Things (IoT) coupled with automation are presented.

1. Introduction

To have a secure future it is very important that modern technologies should be properly analysed in terms of their sustainability aspects. The main reason to practice sustainability is to ensure harmony between humans and the environment we live in. In this economic race to the top, people and governments often concentrate only on the economic gains and disregard the impacts of decisions taken to enhance economic gain on the environment or/and society. Governments all over the world have endorsed initiatives to promote sustainability in all their activities. EPA (environmental protection agency, USA) has implemented initiatives to endorse the executive order 13693 released to reduce greenhouse gas emissions by 40% from the 2008 levels in the next decade [1]. The World Bank has launched many sustainability driven programs such as '*Shared Infrastructure for Solar Parks*' [2] and several projects under '*Global Environment Facility*' (GEF) [3–5] to ensure and promote sustainable development throughout the world. Further, many organizations, academics and researchers around the globe have developed sustainability measurement indices to help us analysing region's sustainability status. Happy Planet Index (HPI) [6], Quality of Life Index (QLI) [7] and Environmental Performance Index (EPI) [8] are few of the well-known sustainability measuring Indices.

IOP Conf. Series: Materials Science and Engineering **244** (2017) 012020 doi:10.1088/1757-899X/244/1/012020

Automation is the one of the current modern innovations, and undoubtedly it can be referred as future's technology. At this stage, it is particularly important to define the framework to assess the sustainability aspect of this technology. Starting with the invention of computers that revolutionized human work, inundated many clerical positions, today, the automation and robotics industry has only become smarter and more complex. We have become so used to automation and no longer realize that most of our interactions are with autonomous machines and automated control systems. Society is evolving to accommodate autonomous machines and this necessitates the study of their impact on our social life.

The aim of this study is to analyse the social effects of automation on the society. Sustainable development is discussed and '*Circles of Sustainability*' approach [9] is further utilized to establish a framework of the current analysis. The social impact is studied by analysing cultural and ethical concerns associated with the automation and reviewing its effect on the human employment opportunities. Further, paper also provides review and discussion on the coupling of other modern technologies with automation such as internet of things (IoT) and outlines the associated future implications.

2. Framework

The World Commission on Environment and Development (WCED) defines sustainable development (SD) as "A process of change in which the exploitation of resources, the direction of investments, the orientation of technological development and institutional change are all in harmony and enhance both current and future potential to meet human needs and aspirations" [10]. To analyse the sustainability of a technology or a system, it requires a huge amount of data and can only be approximated as it is impossible to measure all the dimensions of sustainability simultaneously. Among the many philosophies that provide us a framework for measuring sustainability, three pillars approach also known as triple bottom line (TBL) approach is the most basic and widely accepted. In this study, a new derived form of three pillars approach is used that is termed as '*Circles of Sustainability*'. The approach was developed by Paul James and outlined in the book '*Urban Sustainability in Theory and Practice*' [9].

In literature, the '*Circles of Sustainability*' approach is mainly utilized to assess the urban sustainability profiles of different regions and the outcomes are focused more towards the social aspects of sustainability. Figure 1 shows the schematic illustration of the '*Circles of Sustainability*' approach. The approach assesses sustainability using four dimensions of ecology, economics, politics and culture. Each of the four major domains are further divided into seven subdomains as shown in the Figure 1. In order to examine sustainability of each subdomain a nine point scale has been introduced, where lower point reflects critical sustainability and highest point reflects vibrant sustainability as shown in the legend of Figure 1 [9]. The major difference between the TBL approach and the circles of sustainability approach is the division of social aspects into two weighted parameters of politics and culture. Since this study is more focused towards analysing the social impact of automation, so '*Circles of Sustainability*' approach has been preferred over the TBL method.

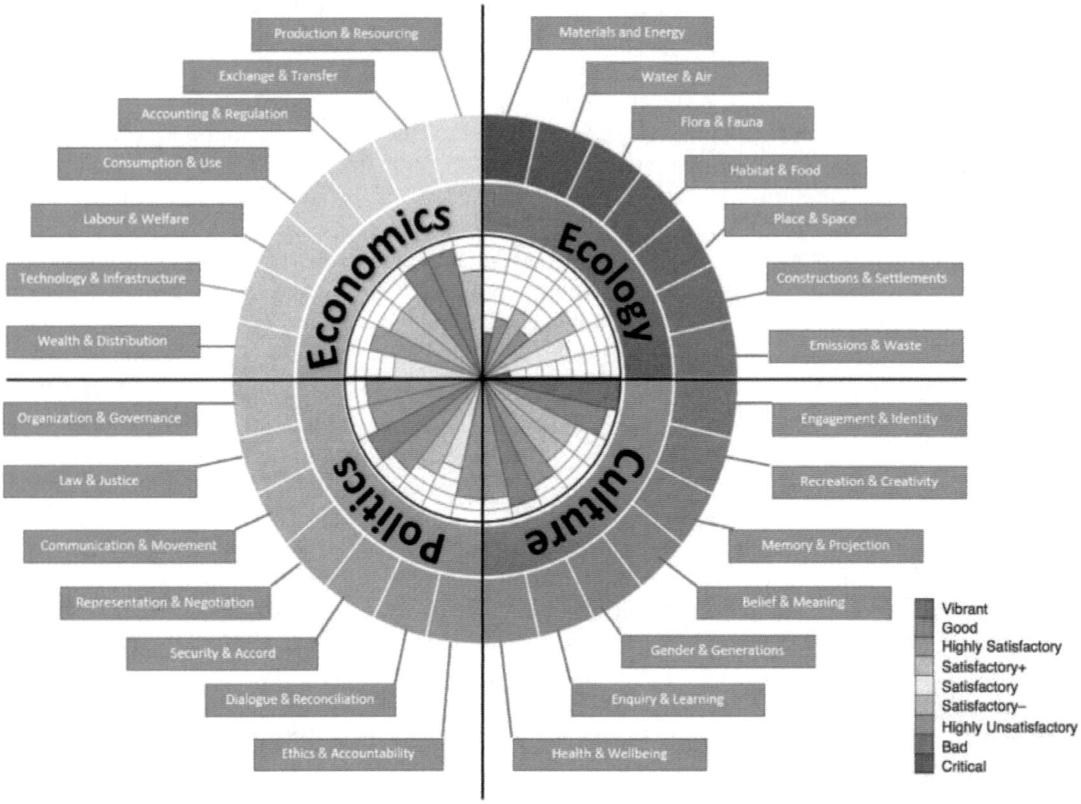

Figure 1. Schematic illustration of Circles of Sustainability Approach (adopted from ref. [9]

3. Synchronizing circles of sustainability approach for automation

Based on the above-mentioned framework, selective subdomains from the major four domains of ecology, economics, culture and politics, have been selected to characterize social implications of automation technology. These subdomains have been selected based on the availability of literature data. The selected subdomains have been represented in the Figure 2 as shown below.

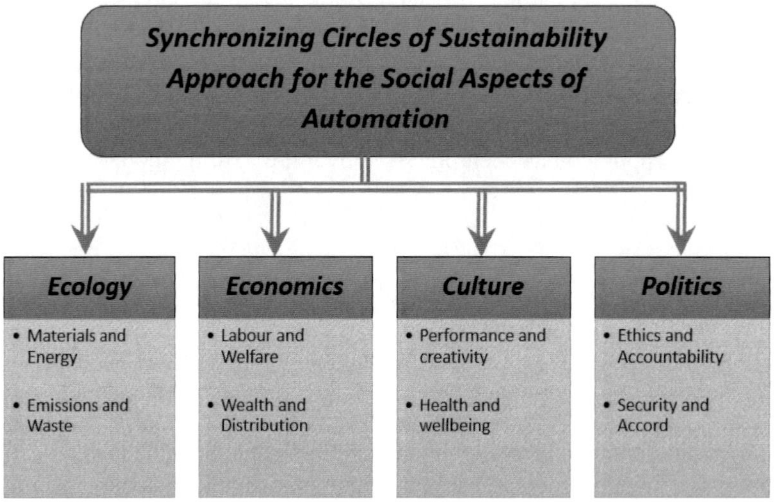

Figure 2. Selected subdomain to reveal the social impact of automation using circles of sustainability approach.

3.1. Ecology
The fact that automation based technologies are the major advancements in science and the most promising technologies of the future, it is imperative that one should take the necessary measures to limit its effect on the environment. These effects are mainly discussed under ecology domain. The literature about the selected subdomains of materials and energy, and emissions and waste has been discussed as under.

3.1.1. Materials and energy
The idea of automation is to maximize the profit margins by increasing the production rate. The idea also supports the economic concern as for the bulk production rate automation reduces the cost as compared to the manual production. At the same time, it should be kept in mind that automation based technologies consumes electricity as input energy and the associated electricity consumption cannot be neglected [11]. It is also well understood that energy consumption in automation should be reduced or optimized to attain sustainable future. Several researchers have focused their work to analyse and optimize the energy consumption in automated based technologies. Vergnano et al. [12] explored the energy optimization in an automated manufacturing system using an embedded approach that incorporates energy evaluation of the system into the scheduling model of the system. The study reveals an enormous energy saving potential using this schedule driven modelling approach. In another study, Paes et al. [13] studied the energy efficient path planning of an industrial automated robot. The study reveals that intelligent energy efficient trajectory planning can save time and energy consumption. The study provided that the most efficient trajectory saved 4% of the energy consumption and 3% of the time. Nilakantan et al. [14] examined the energy optimization potential based on the cycle time of an automated automobile body paint assembly line. The study provided models for the energy balancing using cycle times and revealed good potential.

3.1.2. Emissions and waste
As automation requires electricity as main input source of energy, the associated emissions cannot be neglected when it comes to the sustainability evaluation. As per International Energy Agency (IEA), 24% of the total direct CO_2 emissions came from the industry that appears to be the second largest source of emission [15]. The waste generation due to the automation is very industry specific topic. For example, the waste generation rate is different for an automated automobile industry than the semiconductor based electronic industry. However, when it comes to the waste management, automation technology has provided benefits to deal with the recycling opportunities of the waste.

3.2. Economics
There is no doubt that automation based technologies have provided the economic gains to the industry. However, the economic social subdomains such as labour and welfare, and wealth and distribution have been discussed further.

3.2.1. Labour and welfare
Economic sustenance is important to ensure welfare of a community. Employment is the source of financial input to many families and hence an important criterion to analyse and assess sustainability. Literature on the effect of automation on the employment opportunities was abundant up to late 1980's but since then has been only rarely evaluated. The effect of automation on society has been discussed in view of available literature. The integration of automation technologies into the workplace has changed the job requirements and division of work are no longer the same. The boundary for this change in work division can be best plotted to the introduction of computers to work spaces [16].
Literature is divided on whether the overall effect of automation is negative or positive but there is a consensus that automation is 'job killer' for repetitive simple tasks and 'job creating technology' for everything beyond that [17]. Guteri [18] illustrates examples where the number of people required for certain jobs reduced due to the introduction of automated systems in the Industry. Automation in the logistics and inventory systems has significantly reduced the number of people required to maintain storage and retrieval mechanisms in the respective industries. Further, levels of automation have

continuously evolved leading to the smarter technologies and consequently much lower employment opportunities in that sector [19]. Flexible manufacturing lines also help in performing different tasks with minor modification to assembly lines and provide better productivity [18]. In contradiction to the former claim that automation reduces the number of employees, Frey and Osborne [16] discusses the idea of technology favouring the highly skilled workers is a twentieth century phenomena. In the beginning, automation served the purpose of deskilling activities and splitting them to number of low skill repetitive tasks. Frey and Osborne [16] have conducted a numerical study to analyse how susceptible are jobs to Automation. Three important characteristics on which probability of automation is dependent were identified for the analysis. These characteristics were perception and manipulation, creative intelligence and social Intelligence. An algorithm was developed using O*NET software used by the US department of labour to evaluate 702 job descriptions based on the above three characteristics. Inputs such as finger dexterity, originality for each job were given as inputs to the algorithm and results were presented as high, medium or minimal risk of susceptibility to automation. The results attested some of the claims in literature such as decreased job opportunities in logistics and transportation related sectors whereas the high cognition related jobs such as engineering where in minimal risk category. A surprising result observed in the analysis was that number of jobs in the service sector was classified under high-risk category.
Another prevalent point of view is that the aggregate jobs has remained the same even though there have been changes in the focus area of human labour [20]. The study supporting this claim, states that even though certain job identities have been taken over by automation and robotics, that has alternatively given rise at least equivalent amount of middle level "blue collar" jobs but with a different identity. The study presented in [20] also discusses an example of the advent of motor vehicle. In 1920 due to the advent of motor vehicle the number of equestrian modes of transport slowly became non-existent but at the same time, market was generated for more number of street vendors. In the modern day, job titles such as radiology technicians, phlebotomists are described as the new middle level blue collar jobs. Though the average employment numbers would remain the same, the problem is in identifying where the labour requirement is now focused, as the evaluation of the same is only accurately possible after a certain period. Another aspect associated with new job creation is the educational level associated with the job. It becomes very difficult for working adults to revisit their educational capacities in order to re-qualify for a new job.

3.2.2. Wealth and distribution

An important characteristic of automation technologies is that in most cases these technologies increase productivity, therefore increasing the economic gain. Governments all over the world are pushing industries towards automation to increase the production capacity and improve their economic status. Hence, it is a definite fact that wealth will increase with the introduction of automation technologies [21]. The first industrial revolution witnessed in the UK was a very good example of the economic gains initiated through automation technologies. However, the question of wealth distribution is quite debatable. Even though the cost of automation technologies is reducing, the initial cost is still higher for many entrepreneurs with few resources. Further, since automation has led to deskilling of works, the wages for these works are also reduced. At the same time, high skill jobs such as engineering and design, as a by-product of automation, are associated with higher wages paid, as discussed in [22]. Overall, even though the wealth generation is guaranteed, equal distribution among the members of the society is a definite problem.

3.3. Culture

Keeping in view the cultural domain of sustainability, following subdomains of 'performance and creativity' and 'health and wellbeing' have been discussed under the context of automation.

3.3.1. Performance and creativity

Automation provides better and more efficient way of doing specific tasks but automation hardware and software will take time to replace the human cognitive skills. It means that boundaries of creativity have been extended with the introduction of automation to the society but still it has limitation when it comes to the human cognitive skills for critical thinking [23]. In the current form automation aids extending the limits of creativity but requires human interaction or collaboration.

3.3.2. Health and wellbeing

Health and wellbeing of the people is an important criterion for evaluation and it is impossible to ignore the benefits of automation in this sector. Apart from the many developments in the field of medicine and surgery, automation and robotics have replaced humans in many dangerous job titles such as underwater operations [24]. In harmful and harsh operations such as welding the robots can help the humans in performing the job faster and thereby reducing the operating time of these jobs. Similarly, automated robots can handle harsh environments where human interaction is very difficult such as space exploration, deep ocean exploration, radioactive facilities and toxic environments. In addition, automated robots are also being used to perform the domestic tasks such as cleaning and washing related activities. Automated robots are also being used for the surveillance applications and activities to unsure safety of the community [25]. Further, the entertainment related automation would add to the job market and contribute towards welfare of the society.

3.4. Politics

Keeping in view the political domain of sustainability, following subdomains of ‘ethics and accountability’ and ‘security and accord’ have been discussed under the context of automation.

3.4.1. Ethics and accountability

Healthcare sectors are gearing up to introduce automated robots as primary care personnel at hospitals and clinics. Many robots have already been employed to work as care robots in hospitals. Technology has bred automated robots that can help the elderly to move and cater their important needs [26]. Beran et al. [27] employed robots as companions to children undergoing vaccination. Though the children cried as much as without the robot, the parents attested that the children recalled the robot more than the needle and encouraged clinics to include such distractions. Robots are currently in development to serve functions such as surgical robots, care robots, service robots and in the development of prosthetics [28].

In a study sponsored by the European Commission, the ethical concerns associated with robotics and automation was tackled by applying logical reasoning and experimental study [29]. Ethicists express concern regarding accountability, quality of care, privacy and morality issue associated with the use of robots in the field of healthcare. The cases of liability and supervision in the healthcare field are primary reason for division of opinion regarding robots in healthcare. ‘Who is accountable for the decisions made by the robot’? is the question that ethicists demand answer [29]. Another major issue associated with the use of robots is the quality of care. Could robots truly replace the care one human can provide another? The people opposed to the use of robots have termed robotic care to ‘*Mechanical*’ and missing the warmth of human care. The claim is made on the basis that since robots cannot feel or sense the feelings of its patients, a vital component of human care contributing to a patient’s recovery is missing in robotic care. Robots inability to make moral decisions is another factor highlighted by many ethicists. Decisions are not all mathematical and the lack of moral reasoning in robots is another issue discussed in the ethics of healthcare robotics.

3.4.2. Security & accord

In the subdomain of security and accord, the issues related to the trust and security aspects of automation are addressed. In the study [29] the researchers, through questionnaires, find that people do not trust the robots. In the case of domestic service robots, majority of people do not opt of automated robots due to the lack of trust and false misconceptions that robots will overpower humans. Further, people find the use of robots as a breach of privacy and find it difficult to let robots into their private life. People also fear working with fast, powerful machines as companions since there exists a lack of trust. The study presents the need to educate people about the technology of robotics and clear the misconceptions prevalent in the minds of people. When working with powerful machines, the study recommends appropriate safety must be implemented. Security robots are the future being discussed

by the current world leaders but proper guideline for data protection is cited as a requirement in the study.
Today's world is driven by the internet and almost every service in the world today is made possible through the IoT. The coupling of automation and other modern concepts such as internet of things (IoT) should be discussed when addressing the security related issues of automation. To date, security frameworks for the implementation of IoT across industry verticals is generally lacking. IoT devices have gathered pace at the expense of security best practices in a race to deploy solutions as the first in the market, usually funded by venture capitalists looking for quick return on investment. Areas of focus in the future will need to consider securing the operating systems that the automation tasks are executed on as well as the wireless communications environment needed in the world of automation through IoT.

4. Conclusions and future recommendations

This study is aimed at reviewing the social impacts of automation on our society. The study utilized 'circles of sustainability' approach as a framework to analyse the technology of automation. Following conclusions can be drawn from this current study.

- Automation delivers improved and effective way of doing specific tasks but automation will take time to replace the human cognitive skills for critical thinking. This means creativity is facilitated through automation but requires human interaction.
- Ethical aspects of automation revealed several concerns such as inability to make moral decisions, accountability of medical related decisions, and lack of human feelings.
- It was observed that many life cycle analysis (LCA) studies do not exist on automation. Research must be encouraged to study the environmental effect of automation technologies through LCA analysis to better understand the environmental impact.
- Circles of sustainability has several social subdomains present, however in the current study only few of the subdomains are covered. More detailed studies are required to gather data and conduct complete analysis.

5. References

[1] EPA (environmental protection agency U Executive Order 13693: Planning for Federal Sustainability in the Next Decade *2015*
[2] The World Bank Shared Infrastructure for Solar Parks Project *March 30, 2017*
[3] The World Bank GEF Sustainable Landscape Management Project *March 23, 2017*
[4] The World Bank Zambia Integrated Forest Landscape Project (GEF) *May 4, 2017*
[5] The World Bank GEF Mainstreaming Integrated Water and Environment Management *May 9, 2016*
[6] New Economics Foundation (NEF) Happy Planet Index
[7] Carol Estwing F and Powers M Quality of Life Index (QLI) *1984*
[8] Hsu A et al. 2016 Environmental Performance Index. New Haven, CT: Yale University.
[9] James P, Magee L, Scerri A and Steger M *Urban Sustainability in Theory and Practice Circles of sustainability* (London: Routledge)
[10] The World Commission on Environment and Development 1987 *Report of the World Commission on Environment and Development: Our Common Future*
[11] Bugmann G, Siegel M and Burcin R 2011 A Role for Robotics in Sustainable Development? 13–5
[12] Vergnano A and Thorstensson C 2012 Modeling and Optimization of Energy Consumption in Cooperative Multi-Robot Systems **9** 423–8
[13] Paes K, Dewulf W, Vander K, Kellens K and Slaets P 2014 Energy e ffi cient trajectories for an industrial ABB robot **15** 105–10
[14] Nilakantan J M, Huang G Q and Ponnambalam S G 2015 An investigation on minimizing cycle time and total energy consumption in robotic assembly line systems *J. Clean. Prod.* **90** 311–25
[15] WHEB Commentary Sustainability, automation and the future of manufacturing

[16] Frey C B and Osborne M A 2013 The future of emploment : How susceptible are jobs to computerisation? (available at: http://www.oxfordmartin.ox.ac.uk/downloads/academic/The_Future_of_Employment.pdf) 1–72
[17] Černetic J, Strmčnik S and Brandt D 2002 Revisiting the social impact of automation *IFAC Proc. Vol.* **35** 167–78
[18] Guteri F 1983 An unanswered question: automation's effect on society *IEEE Spectr.* **20** 89–92
[19] Vagia M, Transeth A A and Fjerdingen S A 2015 Review article A literature review on the levels of automation during the years . What are the different taxonomies that have been proposed ? *Appl. Ergon.*
[20] Autor D H 2015 Why Are There Still So Many Jobs? The History and Future of Workplace Automation *J. Econ. Perspect.* **29** 3–30
[21] Azzone G and Bertele U 1989 Measuring the economic effectiveness of flexible automation: a new approach *Int. J. Prod. Res.* **27** 735–46
[22] Chui M, Manyika J M M 2015 Four fundamentals of workplace automation. *McKinsey Quarterly, Novemb.*
[23] Pinder R 2017 How automation could make us more human (Available at: https://www.campaignlive.co.uk/article/automation-us-human/1428520)
[24] Kelasidi E, Pettersen K Y and Gravdahl J T 2015 Energy Energy efficiency efficiency of of underwater underwater robots *IFAC-PapersOnLine* **48** 152–9
[25] Gupta V and Bala R 2013 Advancements in Robots for Human Welfare : A Review *Int. J. Res. Manag. Sci. Technol.* **1** 106–15
[26] Holder C, Khurana V, Harrison F and Jacobs L 2016 Robotics and law: Key legal and regulatory implications of the robotics age (Part I of II) *Comput. Law Secur. Rev.* **32** 383–402
[27] Beran T N, Ramirez-Serrano A, Vanderkooi O G and Kuhn S 2013 Humanoid robotics in health care: An exploration of children's and parents' emotional reactions *J. Health Psychol.* **20** 984–9
[28] Bogue R 2011 Robots in healthcare *Ind. Robot An Int. J.* **38** 218–23
[29] Leroux C and Labruto R 2012 EU Robotics D3.2.1 Ethical Legal and Societal issues in robotics 1–38

Study on vacuum packaging reliability of micromachined quartz tuning fork gyroscopes

Maoyan Fan[1] and Lifang Zhang[2]

1 School of physics and Electronic Engineering, Yuxi Normal University, Yuxi 653100, Yunan Province, China
2 Office of Academic Affairs, Yuxi Normal University, Yuxi 653100, Yunan Province, China

Abstract. Packaging technology of the micromachined quartz tuning fork gyroscopes by vacuum welding has been experimentally studied. The performance of quartz tuning fork is influenced by the encapsulation shell, encapsulation method and fixation of forks. Alloy solder thick film is widely used in the package to avoid the damage of the chip structure by the heat resistance and hot temperature, and this can improve the device performance and welding reliability. The results show that the bases and the lids plated with gold and nickel can significantly improve the airtightness and reliability of the vacuum package. Vacuum packaging is an effective method to reduce the vibration damping, improve the quality factor and further enhance the performance. The threshold can be improved nearly by 10 times.

1. Introduction

Currently the design, processing and manufacturing technologies of the microelectromechanical systems (MEMS) are developing rapidly, but the development and application of MEMS devices are restricted because the packaging technology lags behind the processing technology. The service life of vacuum packaged MEMS devices depend on the vacuum level of the device cavity and the vacuum hold time. The real-time monitoring of the vacuum level in the cavity is a key technique for developing MEMS devices [1]. The packaging process of micromachined quartz tuning fork gyroscopes comprises the wire bonding, die attachment and seal cap. In order to ensure the packaging reliability of micromachined quartz tuning fork gyroscopes, the damping to the fork resonance caused by the air or inert gasses in the cavity must be reduced, so as to improve the sensitivity and comprehensive performance of gyroscopes. Even packaged in a high vacuum environment, the comprehensive performance of quartz tuning fork gyroscopes is not directly proportional to the vacuum level. If the damping is too small, "oscillation" occurs. Selecting an appropriate vacuum packaging process is important for isolating the effects of external interfering signals and obtain the highest quality factor (Q-factor), sensitivity and comprehensive accuracy [2, 3].

2. Bonding technology of micromachined quartz tuning fork gyroscop

Packaging is a key process for the micromachined quartz tuning fork gyroscopes. The performance of gyroscope is influenced by the encapsulation shell, encapsulation method and fixation of forks. The quartz tuning fork is Z-cut crystal quartz wafer with a thickness of hundreds of microns, and its fabrication processes include grinding and polishing, photolithography, etching, evaporation coating of silver film electrodes, and packaging. Maintaining high vacuum for a long period of time is a necessary measure to control the noise level of micromachined quartz tuning fork gyroscopes, which can directly affect the working stability, service life and shock resistance of quartz tuning fork. The single-point fixed packaging is good for stabilizing the working performance of quartz tuning fork. Main considerations for selecting the shell materials are elaborated as follows. (1) The iron-nickel alloy is one commonly used material, which comprehensively takes account of the strength, coefficient of thermal expansion (CTE) and airtightness. (2) The vacuum packaging process is utilized and the structure and dimension of the resonant cavity are determined according to the quartz tuning fork size. The tuning fork used in the experiment is $12mm$ long, $2.5mm$ wide and $0.4mm$ thick. The encapsulation shell base is $20mm$ long, $10mm$ wide and $8mm$ high and its size satisfies the packaging requirements of the forks. It is encapsulated with the vacuum level of $\leq 1\times10^{-2} Pa\cdot m^3/s$. Refer to Table.1 for the material properties.

Table.1. Material properties of vacuum package shell.

Material	Density ρ/gcm-3	Shear Modulus G/Gpa	Poisson's ratio μ	Young's modulus E/Gpa
iron-nickel alloy	8.1	79.4	0.3	206.0

The quartz tuning fork must be fixed to the encapsulation shell base after being processed. Common fixed packaging methods include single-point or two-point package. The latter method is employed for the quartz tuning fork. Because the quartz and shell have different CTEs, the stresses generated from external vibrations and temperature protection are prone to affect the tuning fork performance. The performance of the single-point package of resonant nodes of quartz tuning fork is better than the two-point package. The structure is shown in Figure 1. The nickel and gold films are plated on the bonding regions of the quartz tuning fork by means of sputtering, evaporation or electroplating, and the anodic bonding is achieved by using the vacuum packaging machine [4]. The gold-plated bonding region of quartz tuning fork is attached to the convex surface of the encapsulation shell base and placed on the worktable in the vacuum furnace. A certain pressure is applied. When the vacuum and heating reach the set values, apply a voltage of $800V_{DC}$, the bonding region of quartz tuning fork is connected to the high voltage anode and the shell base connects with the cathode. Heat the bonding region to the gold-silicon eutectic temperature $(363^{\circ}C)$. The gold-silicon inter-diffuses on the bonding interface and forms the eutectic mixture. When the temperature exceeds the eutectic point, a strong eutectic bonding layer is shaped as the time increases. The electrodes of quartz tuning forks are welded by gold wires with the encapsulation pin through the gold wire ball bonding or hot bonding process. The gold wire ball bonding is characterized by the big contact surface, corrosion resistance, high tensile strength, good reliability and easy automatic or semi-automatic operation [5,6].

Figure 1. Quartz tuning fork by single-H single-point package.

The Q-factor of quartz tuning fork is related not only to the quartz material, electrode distribution, tine size, and shape, but also to the packaging vacuum level of the tuning fork. If the Q-factor is very high, the damping coefficient can be used for assessing the quality factor of tuning forks. The Q-value and the damping coefficient α have the following approximate relationship: $Q \cdot \alpha = 1/2$. It is packaged in an atmospheric environment and in the vacuum state. The impedance analysis test is conducted for the drive tine of quartz tuning forks. The main test data include the equivalent motional capacitance R_1, equivalent motional inductance L_1, equivalent shunt capacitance C_0, equivalent motional capacitance C_1 and resonant frequency f. The quality factor Q may be calculated by the equation below:

$$Q = \frac{1}{2\pi f C_1 R_1} \tag{1}$$

The damping coefficient α can be obtained by Eq.(1) after calculating the quality factor Q. When the sample packaging experiment is conducted in an atmospheric environment, the lower the quality factor Q of the quartz tuning fork, the larger the damping coefficient α, and the motional capacitance R_1 reflecting the mechanical loss also becomes larger. The resonant frequency of the quartz tuning fork can be calculated through the equivalent motional capacitance C_1 and the equivalent motional inductance L_1:

$$f = \frac{1}{2\pi\sqrt{L_1 C_1}} \tag{2}$$

The experimental results show that the damping coefficient becomes larger, the equivalent motional capacitance becomes smaller, the equivalent motional inductance becomes larger and the resonant frequency of the tuning fork decreases. The vacuum level has a significant impact on the performance of the quartz tuning fork. It can reduce the damping coefficient and improve the quality factor. Therefore, if the vacuum package is employed, the damping coefficient of gasses in the shell is greatly reduced and the quality factor of the micromachined quartz tuning fork gyroscope can be significantly improved.

In the packaging process of micromachined quartz tuning fork gyroscopes, the quartz tuning fork must first be attached to the convex support of the shell substrate by the welding or bonding process. The bonding technology generally adopts the integral heating, which is time-consuming. The high temperature is easy to cause thermal damage to the sensitive element and circuit. The mismatch of the CTEs of the two kinds of bonding materials leads to the increasing thermal stress in the bonding region as well as the poor device reliability. In order to reduce the adverse factors of thermal bonding, two technical approaches are usually taken: first, the low-temperature solder and low-temperature bonding technology; second, the localized induction heating packaging technology, for the package of devices containing temperature sensitive chip and structure when the temperature difference is greater than 200 ℃. It can be sealed in the air or achieve the vacuum hermetic package at a low cost and the localized heating of the electromagnetic induction bonding region. The microsystem electromagnetic induction local-heating packaging equipment is shown in Figure 2.

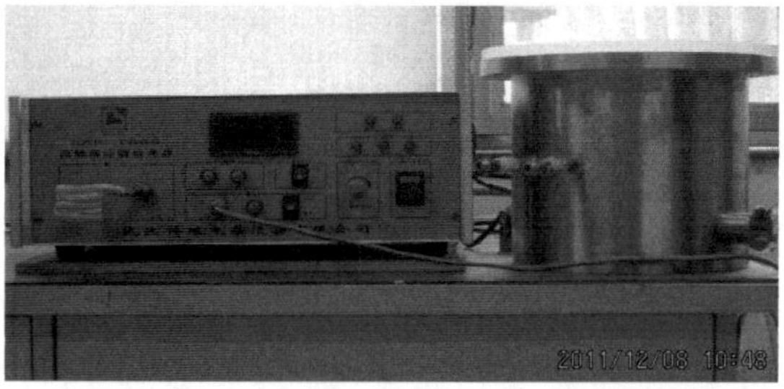

Figure 2. Electromagnetic induction localized-heating and bonding packaging equipment.

3. Vacuum packaging process of micromachined quartz tuning fork gyroscopes

The shell of micromachined quartz tuning fork gyroscope can adsorb gasses. The airtight materials in a high-temperature environment release the adsorbed gasses and thus affecting the packaging vacuum level. Before packaging, adsorbed gasses of the shell must be removed to ensure the reliability of the gyroscope vacuum packaging. Experiments show that the average vacuum level of the unbaked shell is $9 \sim 10Pa$ after vacuum package, but that of the baked device is about $5Pa$. The vacuum level is nearly doubled and beneficial to improve the gyroscope precision [7].

In the case of good welding quality, the shell leakage mode is the approximate molecular flow mode. The vacuum cavity volume is V and the vacuum pressure of the vacuum cavity of encapsulation shell is P. Assume that the air pressure is P_{air}, the vacuum pressure in the buffer cavity is P_1 and the buffer cavity volume is V_1. Gasses may only pass the outside of the sealed cavity through the first welding rib into the vacuum buffer cavity, where the gas pressure rises. Then gasses pass through the second welding rib into the vacuum cavity. According to the molecular flow theory, if the leakage permeance is identical for inner and outer welding ribs, assume that the air leakage permeance is C, and the equation is as follows:

$$V_1 \frac{dP_1}{dt} = C(P_{air} - P_1) - C(P_1 - P) \tag{3}$$

$$V \frac{dP}{dt} = C(P_1 - P) \tag{4}$$

Assume that the pressure in the initial buffer cavity and the vacuum cavity is P_0 and solve the two equations, we obtain the change of vacuum cavity pressure P with time:

$$P = \frac{r_2(P_{air} - P_0)}{(r_1 - r_2)} e^{r_1 t} - \frac{r_1(P_{air} - P_0)}{(r_1 - r_2)} e^{r_2 t} + P_{air} \tag{5}$$

Meanwhile, the change of pressure in the buffer cavity with time can be obtained:

$$P_1 = \frac{(P_{air} - P_0)}{(r_1 - r_2)} \left(\frac{C}{V_1} + r_2 \right) e^{r_1 t} - \frac{r_1(P_{air} - P_0)}{(r_1 - r_2)} \left(\frac{C}{V_1} + r_1 \right) e^{r_2 t} + P_{air} \tag{6}$$

Where: $r_1 = \frac{-b + \sqrt{b^2 - 4c}}{2}; r_2 = \frac{-b - \sqrt{b^2 - 4c}}{2}$; $b = \frac{2(CV + CV_1)}{V \cdot V_1}; c = \frac{C^2}{V \cdot V_1}$

The encapsulation shell has no buffer cavity suitable for the equation:

$$V \frac{dP}{dt} = C(P_{air} - P) \tag{7}$$

The change of pressure with time is:

$$P = P_{air} - (P_{air} - P_0) e^{-\frac{C}{V}t} \tag{8}$$

After improving the welding quality, the leak rate usually reaches $10^{-9} \sim 10^{-13} Pa \cdot m^3 / s$. Assume that the vacuum failure criterion is the change of vacuum level by 50%. By using Eqs.(5) and (8), a rough calculation of the vacuum hold time is obtained for the vacuum package with and without buffer cavity, at least extended by over 20 times. Generally, the vacuum packaged devices are required to hold for more than 8 years. The existing getter has limitations in its application in the vacuum package of micromachined quartz tuning fork gyroscopes. The encapsulation shell with buffer cavity can reduce the leak rate. Its structural design is shown in Figure 3. The key to the design of the encapsulation shell is to ensure that all the gas leakage pathways pass through the buffer cavity, the shell buffer cavity and the vacuum cavity must have the same initial vacuum while being packaged, and the buffer cavity is directly formed by the resistance welding of the inner and outer welding ribs [8, 9].

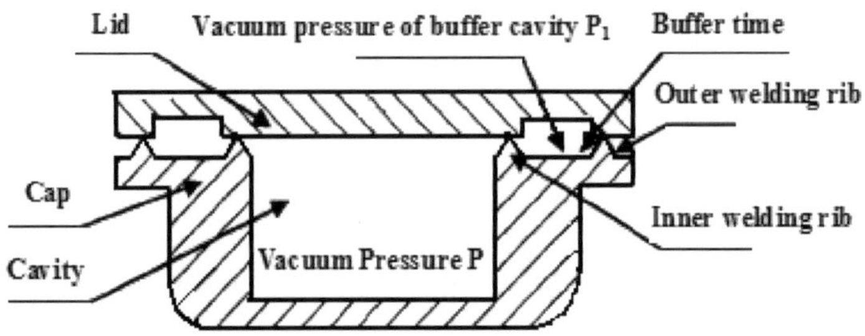

Figure 3. Design of vacuum package shell with vacuum buffer cavity.

According to the welding process characteristics, at the squeeze stage the shell cap is in close contact with the base and then the welding current is connected. Due to the restriction in the accuracy of the shell fabrication process, the roughness is not uniform for the contact surface between the seal cap and the base, resulting in too large contact resistance at the welding position. The excessive joule heat will burn the contact surface when the current passes, so the preliminary squeeze is needed. If without sufficient squeeze or pressurization, the uneven current may lead to a lot of solder splash or local burn. After the squeeze stage, the passage of current generates heat, and plastic deformation occurs with broken grains at the interface under pressure and heat. The broken grains are recrystallized at a high temperature so that common crystal grains are formed between weld pieces through the mutual crystallization to complete the welding in the plastic state. As the temperature rises, the plastic area extends around and there is a melting area in the center. Since the melting area is surrounded by the plastic area, the liquid molten core will not splash under pressure and the molten solder will not be oxidized by external air. However, if the temperature is too high, the melting area expands to the edges of the weld seam and causes an outflow of the molten metal. The gold-tin solder is molten, then the power is cut off and the welding pressure is held. The molten gold-tin solder gradually cools under the normal temperature and becomes solidified and crystallized to form alloy. The rapid cooling of liquid gold-tin solder must be avoided to prevent the formation of a brittle alloy structure and a small strength at the welding position that is brittle and especially easy to leak.

In order to improve the weld strength, the secondary pulse method can be used to change the brittle alloy structure. With the secondary pulse current, the gold-tin solder alloy is directly annealed into the void-free alloy structure with good ductility and fine grains. The weld strength can achieve satisfactory results after good control of the annealing current and time, adjustment of the parameters of welding pressure, current and time according to the secondary pulse process, and consideration of other factors such as weld strength. The packaging process of micromachined quartz tuning fork gyroscopes directly affects the vacuum level inside the shell, thus influencing the gyroscope performance. In order to optimize the vacuum package process of micromachined quartz tuning fork gyroscopes, the process parameter optimization test of the vacuum end cap by using the empty shell is employed to test the vacuum package reliability. It is tested on the vacuum packaging machine developed by Huazhong University of Science and Technology. The vacuum seal cap utilizes the reflow soldering of Au-Sn solder with the process as follows:

(1) The Au-Sn plated base and shell for the package of micromachined quartz tuning fork gyroscopes are cleaned for 10 minutes by using the ultrasonic acetone, so as to remove the oil dirt and ensure the surface cleanliness of the package;

(2) The packaging material and gasses adsorbed on the component surface are removed and it is baked in an 80℃ vacuum drying oven for 8h;
(3) The package base, solder material, and seal cap are placed in the special clamp of the vacuum packaging machine. The vacuum extraction system on the vacuum packaging machine is started and vacuumed to $10^{-2} Pa$ so as to observe the vacuum level. The optimized parameters are used for resistance welding and the vacuum is held for a period of time;
(4) The packaged device is taken out when it cools off to the normal temperature, and the gross leak detection and the fine leak detection of the packaged shell and welding quality are carried out.
Six groups of optimization experiments using different process parameters are designed for the packaging process of micromachined quartz tuning fork gyroscopes. The metal ring on the Kovar metal base and shell has good flatness and density. Its surface must be plated with nickel and gold-tin and the thickness should be greater than $1.25 \mu m$.
The factors influencing the airtightness of the vacuum seal cap of micromachined quartz tuning fork gyroscopes during the optimization of its process parameters are as follows:

(1) The applied pressure is not uniform, the shell is deformed, the hot molten gold-tin solder flows between the seal cap and the shell base and the solder is deficient in local areas, so the seal-tightness is poor and the welding strength is bad.
(2) The applied pressure is too large, the gold-tin solder flows out of the welding position after squeeze and thus the shell appearance is affected. The gold-tin solder is attached to the inside and outside of the shell and leads to the wire short circuit and device failure.
(3) The welding temperature is too high for the package of micromachined quartz tuning fork gyroscopes, the surface tension of the gold-tin solder becomes weaker, it is boiling between the seal cap and the base, and holes easily form in the weld layer and affect the airtightness of shell after vacuum package. Cracks easily develop in the holes of the weld layer and shorten the gyroscope's service life.

The interface and welding effect in the vacuum resistance welding area of the micromachined quartz tuning fork gyroscope are shown in Figure 4, where the gold in the Kovar base and shell spreads towards the solder layer and forms a layer of dense eutectic mixtures between the Kovar alloy base and the shell solder layer. There is no crack in the weld interface layer, so the airtightness can be maintained for long.

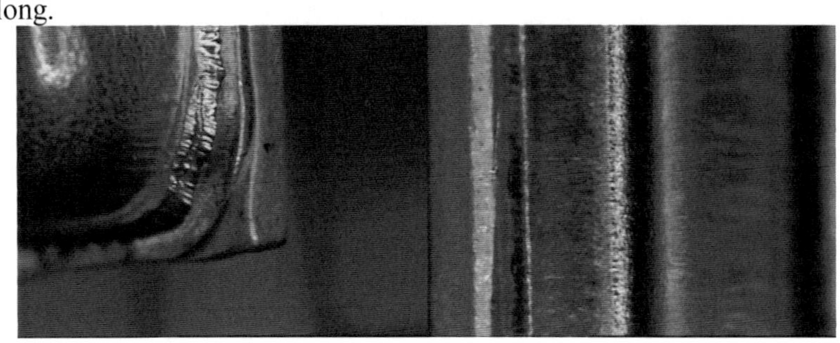

Figure 4. Welding effects at different vacuum resistance welding positions.

4. Testing and analysis of packaging effects

Aetrium model 6000 bubble leak detector is used for the gross leak detection of airtightness of the packaged micromachined quartz tuning fork gyroscopes. The perfluorotributylamine solution (FC-43, boiling point: 165-185℃) is poured into the leak detector and heated to about 90℃. Then the gyroscope resonant cavity is immersed in the solution. Observe from the observation window of the high magnifying glass if there are bubbles released around the weld seam. The gross leak detector and the experimental effects are shown Figures 5(a) and (b). The experimental results indicate no bubbles at the weld seam. Compared with previous experimental data, it was demonstrated that the vacuum

packaged micromachined gyroscope maintains a relatively stable output value after five years, the vacuum package has minor leaks and the airtightness of the devices is kept well [10]. Due to the limitation of experimental conditions, the fine leak detection of the sample hasn't been completed. A helium mass spectrometer is used for the fine leak detection. In this process, the packaged micro-gyroscope device is placed into a pressurization device filled with helium gas for a period of time. Then the leak rate is detected by putting it in a leak rate detection cavity connected with the helium mass spectrometer [11].
The solder material selection is very important for the miniaturization of micromachined quartz tuning fork gyroscopes and high-precision and high-reliability welding pad on the welding surface micro-area of vacuum package. The thermal conductivity is 57w/m·K for the eutectic $AuSn_2O$ with the highest thermal conductivity among solders. The gold content in the eutectic $AuSn_2O$ is 80wt% and the eutectic point is 280℃, so it is an ideal vacuum resistance welding material without the need of flux. The vacuum resistance welding process is, in essence, the reasonable match of three factors, namely pressure, current and time, so as to melt and cool off the gold-tin solder to be alloy. Different heating and melting rates and cooling rates form different alloy structures with big differences in the weld strength. The welding effects indicate that:

(1) The vacuum resistance welding needs no flux and the mechanical strength is high. Gold accounts for a large proportion of the alloy (80%), and the surface oxidation of the material is low. The yield strength of the gold-tin alloy is very high. Even at the temperature of 250~260℃, its strength can meet the airtightness requirements.
(2) The vacuum resistance welding has a low viscosity and a good wettability. The liquid gold-tin alloy has a very low viscosity, thus being able to fill the gap generated by the welding process and having good density.

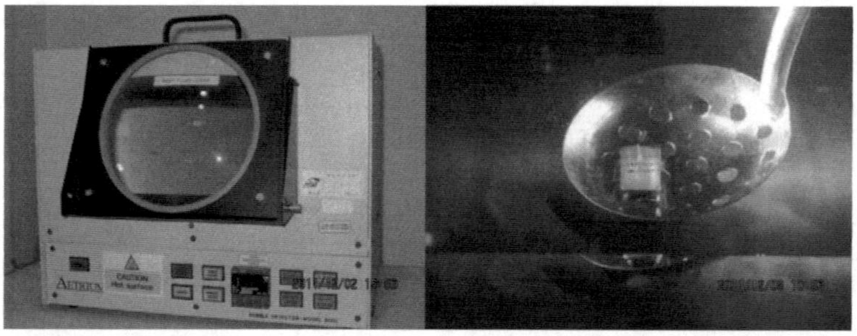

Figure 5. (a) gross leak detector (b) no bubble indicated by the gross leak detection.

After the quartz tuning fork is packaged, if the drive circuit is the same as the sense circuit, the change of gyroscope threshold is only related to the damping change. The damping is determined by the airtightness of the vacuum cavity. The detection of threshold change of micromachined quartz tuning fork gyroscopes can be used to characterize the vacuum package airtightness[12]. The threshold is generally smaller than $0.01^\circ / s$ and can be determined by measuring the relationship of the bias voltage output and the angular rate input. The measurement method is measuring the noise after the gyroscope starts and outputs stably. From the input rate of $0.001^\circ / s$, with an increment of $0.001^\circ / s$, the output curve is recorded and the mean is calculated. When the noise to signal ratio of the output value reaches 63%, the input angular rate is the identifiable sensitivity limit, called threshold. Figure 6 shows the output curve of one minute when the input rate of the micromachined quartz tuning fork gyroscope is $0.005^\circ / s$. It can be determined that the threshold is $0.005^\circ / s$. The tuning fork gyroscope has no dead band caused by friction or other reasons, so the relationship between threshold and noise output is especially obvious. The main source of irregular fluctuation noise is electrical noise, which is caused by the high gain operational amplifier. Another source of noise is generated by the drive motion coupling in output. This effect can be minimized if the gyroscope works at the mechanical zero point by the frequency modulation process.

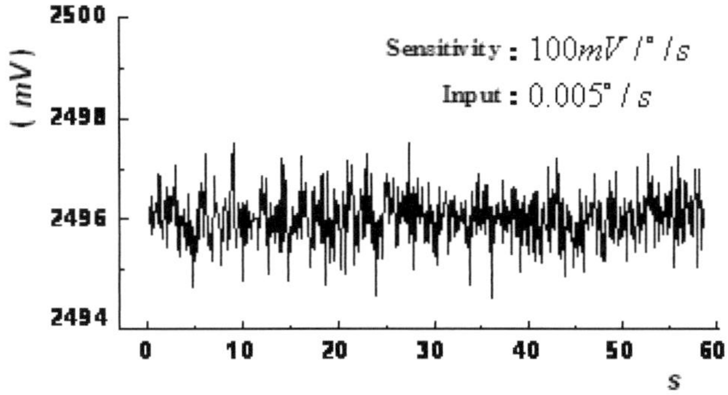

Figure 6. Threshold test of micromachined quartz tuning fork gyroscopes.

5. Conclusion

The vacuum level of micromachined quartz tuning fork gyroscopes should be kept at $1Pa$ to meet the requirements of long-term stability of quartz tuning fork resonance. Resistance welding has the advantages of high yield, short welding time and less material outgassing during welding. The welding surface of alloy solder is characterized by high strength, small thermal resistance, and good stability. The source of gasses in the sealed cavity are mainly gasses leaked from the cavity or released by materials in the vacuum cavity. The internal outgassing is primary compared with the external leakage. The processing of shell material and alloy solder is a key technology for keeping the airtightness. Real-time monitoring of the vacuum level inside small device cavities is a technical problem. Due to the individual differences of quartz tuning forks, the vacuum measurement system outputs different values and needs to be calibrated before online measurement. The online monitoring of resonant resistance of the quartz tuning fork in the small device cavity and the damping coefficient of gasses are related to the measurement of vacuum level, which is of great significance for the optimization of the packaging process of micromachined quartz tuning fork gyroscopes.

6. References

[1] Wang Shanshan, Song Liang, Dang Xiangting. Study on vacuum measurement of the packaged cavity of MEMS device based on QMEMS quartz oscillator [J]. Foreign Electronic Measurement Technology, 2014, 33(1):27-31

[2] J. Wei, H. Xie, C.K. Wong, et al. Nano-film assisted anodic bonding [J]. Proc. of SPIE.2000. 4936:455-463

[3] A. An, Y. Oh, S. Choi et al. Dual-axis microgyroscope with closed-loop detection [J]. Sensors and Actuators, 1999, A:73:1-5

[4] CHEN Bo, DING Rongzheng, MING Xuefei, GAO Nayan. Research on Void Rate of Eutectic Solder Welding [J]. Electronics and Packaging, 2012(12)11:9-12

[5] S. Murali, N. Srikanth, Charles J. Vath III. An Analysis of Intermetallics Formation of Gold and Copper Ball Bonding on Thermal Aging [J]. Materials Research Bulletin.2003, 38(2): 637-646

[6] Zhiyin Gan. Study of vacuum packaging of MEMS devices [D]. Huazhong University of Science and Technology, 2008

[7] Ashwin A. Seshi, Moorthi Palaniapan, Trey A. Roessig, et al. A Vacuum Packaged Surface Micro- machined Resonant Accelerometer. Journal of Microelectromechanical Systems, 2002,11(6):784-793

[8] S.H. Choa. Reliability of vacuum packaged MEMS gyroscopes [J]. Microelectronics Reliability, 2005(45):361-369

[9] Wang Xuefang, Lin Dong, Gan Zhiyin, et al. Research on Vacuum Packaging Process, Equipment and Vacuum Degree Detection of MEMS Devices [J]. China Mechanical Engineering, 2007,18(8):887-890
[10] Wang Chenggang. Investigation of Device Level MEMS Packaging based on Vacuum Resistance Welding Candidate [D]. Huazhong University of Science and Technology, 2013
[11] CHEN Bo, DING Rongzheng, MING Xuefei, et al. Research on Void Rate of Eutectic Solder Welding [J]. Electronics and Packaging, 2012(12)11:9-12
[12] SHEN Qiang, LI Xiaoying, CHANG Honglong, et al. An Effective Method for Improving Mechanical Stability of Vacuum Silicon Micro-Gyroscope [J]. Nanotechnology and Precision Engineering, 2012(10)1:14-19

IOP Conf. Series: Materials Science and Engineering **244** (2017) 012022 doi:10.1088/1757-899X/244/1/012022

Computer aided system for parametric design of combination die

Vishal G Naranje[1], H M A Hussein[2] and S Kumar[3]

1 Birla Institute of Technology, Pilani, Department of Mechanical Engineering Dubai Campus 345055 UAE.
2 Mechanical Engineering Department, Faculty of Engineering, Helwan University, Cairo 11792, Egypt
3 Mechanical Engineering Department, SVNIT, Surat-395007, India
Email: vnaranje@dubai.bits-pilani.ac.in

Abstract. In this paper, a computer aided system for parametric design of combination dies is presented. The system is developed using knowledge based system technique of artificial intelligence. The system is capable to design combination dies for production of sheet metal parts having punching and cupping operations. The system is coded in Visual Basic and interfaced with AutoCAD software. The low cost of the proposed system will help die designers of small and medium scale sheet metal industries for design of combination dies for similar type of products. The proposed system is capable to reduce design time and efforts of die designers for design of combination dies.

1. Introduction

Combination dies are used for production of sheet metal parts having two or more operations in a single station. The terms compound and combination dies have frequently been interchangeably used to define one-station die. Compound dies are used for production of sheet metal parts having combined cutting operations like blanking and piercing, while combination dies are used where two or more sheet metal operations such as forming, drawing, extruding, embossing etc. are combined with each other or with the various cutting operations such as blanking, piercing, trimming, broaching, and parting off. Combination die (Figure 1) consists of several components including die block, die gages, stripper, stripper plate, punch (es), punch plate, back plate, blankholder, die-set and fasteners.

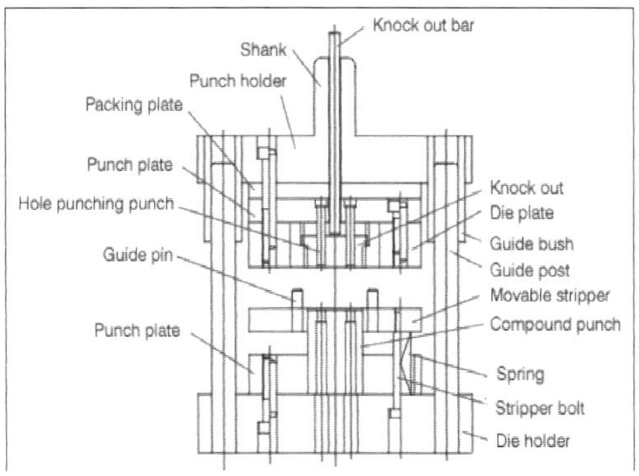

Figure 1. Typical Combination Die. [1]

Design of combination dies is tedious, time consuming and skilled based activity. It requires highly experienced die designers and process planners. Many CAD softwares are available in market to aid die designers for design of dies. But these softwares does not have special die-design functions and capabilities. Furthermore, die designers have to use multiple software packages to perform various tasks of die design process. Also connectivity between these softwares is mostly non-existent. Further, because of high cost of these software packages, very few small scale industries can afford these. Also to operate these softwares and interpret the results generated by softwares well trained competent and experienced designers are required. These sofwares are only useful to improve design efficiency to some extent. Since 1970's and till date, researchers made a lot of efforts to develop computer aided die design systems for design of various types of dies used for production of sheet metal parts. A number of computer aided systems have been developed for punching, blanking, progressive, and deep drawing dies. Very few researchers tried to develop computer aided die design system for compound dies, and fewer research cover the combination die type. Therefore, there is a need to develop an automated system for quick design of combination die. Several researchers applied their efforts in developing computer aided design systems (CADD) for piercing and blanking die [2-6]. Many other researchers successfully developed CADD, knowledge based systems, Expert systems to automate the various tasks of progressive die design, such as [7-9] developed the system for strip layout, punch recognition, punch arrangement and complete design of the progressive die. Further worldwide researchers have also applied their efforts to develop automated systems for deep drawing die [10-13] and compound die [14-19], [20-22]. The foregoing literature review reveals that very few research efforts are found in the area of combination dies [23-27]. In this paper, a knowledge based computer aided system is presented for parametric design of combination die for cupping and piercing operations. The system is flexible and has low cost of implementation.

2. Design considerations

The classification of combinations die depends upon the types sheet metal operations accomplished in one stroke of the press [28]. Figure 2, shows a sample of the expected different types of combination dies from its individual operations.

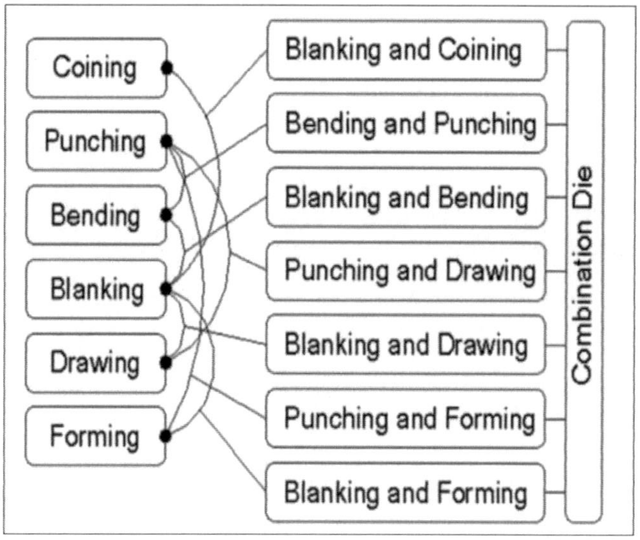

Figure 2. Classification of combination die.

The design, manufacturing and maintenance of these dies are very complicated. Traditionally design of combination dies is based on thumb rules. It involves number of tasks such as determination of cutting force, clearance between punch and die, thrust force, die clearance, force calculation, press tonnage requirement, determination of tool shut height, die block design, design of stripper plate, selection of stripper spring and stripper bolt, design of punch back plate, design of punch holder, design of punch holder, punch design, overall punch length, critical bucking force and critical length of the punch, compressive strength and deflection of the punch, design of guide posts and guide bushings etc. In most of the sheet metal industries, process planners and die designers use their rich industrial experience and knowledge to complete tedious and time consuming task of combination die design. Some of the important basic guidelines generally used for design of combination die are given as under:

1. The minimum thickness used for production of sheet metal parts in compound and combination dies is generally less than 0.4 mm.
2. In combination die for single stage drawing operation the sheet thickness should not be more than 1 mm.
3. All the design features that can be produced without danger or deformation should be pierced prior to wipe bending and forming.
4. Backup plates are made of hardened steel of 3/8 inch (9.5 mm) thick for general work, 1/2 inch (12.5 mm) thick for heavy duty jobs. The hardness should be in the vicinity of 40 to 50 HRC.
5. In the case of combined tool, spring loaded stripping system is used to accommodate bending operations.
6. The design of the die block basically depends up on the work piece size and thickness, the contour of the work piece and the die material.
7. Tool shut height should be kept 5-10 mm less than press shut height to provide a little height adjustment during press setting.
8. Compression type die springs are available in various service grades with corresponding permissible deflection ranges from 25-50% of the free length.
9. To prevent the bulging of the thin stock in combination die a channel type back guides in combination with front spacer and strip is used.
10. To perform two operations a ball bearing die-sets assembly, with guide-post sliding inside a ball bearing which is contained in guide bushing is recommended.
11. It is recommended that shank size should be selected according to the press tonnage and it should be made of medium strength steels with a hardness of 40 - 42 HRC.

Keeping in view of the above basic guidelines and recommendations, a computer aided system labelled as CADCD (Computer aided system for design of combination die) has been developed for parametric design of combination die for cupping and punching operations.

3. Development procedure of proposed system

The proposed system consists of two modules. First module is developed for determination of various process parameters. In the second module a rule based system is used for parametric design of combination dies for piercing and cupping operations. The system includes database that contains data of the mechanical properties of materials, die-set details etc. Initially system invites the user to enter part data information such as production quantity, tolerance, sheet thickness etc. through GUI at appropriate stages during a consultation. User interface of the proposed system is developed using Visual Basic 6.0 interfaced with AutoCAD software. In the first module, die design calculations executes for the die components dimensions. It is based on the relationships between the die components. Second module is developed to model the die components and die assembly of combination die for cupping and punching operations. The module takes required inputs in the form of size of die block, stripper fasteners etc. automatically from output data files generated during the execution of first module. It also takes the inputs from the die component database and part information data base. The proposed system has more than 200 production rules. These rules are coded in Visual Basics 6.0 and interface with AutoCAD. The design rules related to the cupping and punching die, can be simply classified to cutting rules and drawing rules. A sample of production rules incorporated in the proposed system is given in Table 1.

Table 1. A Sample of Production Rules Incorporated in the proposed System.

S. No.	IF	Then
1	Sheet material = Tin/Copper Brass/Stainless steel /Aluminum; and Design feature = Circular/Square/Rectangular/Oblong hole; and Minimum hole diameter/dimension in mm < 0.5	Set minimum diameter/dimension of circular / rectangular/square/oblong hole in mm = 0.5
2	Sheet material = Stainless steel [AISI 1090]; and Shear strength = 600-650 N/mm2; and Sheet thickness in mm > 3.0; and Sheet thickness in mm ≤ 4.0; and Hardness of sheet material = 50-55 HRC	Set clearance all around = 6.4 % of sheet thickness
3	Sheet metal operations= Piercing and drawing and material =Aluminum	Shearing Force = Ls x σs x T Where, Ls - Sheared length, mm; σs - Shear strength, N/mm2; T - Material Thickness, mm.
4	Draw Stage = First; and Diameter of first stage (di) Shell diameter (d)	Drawing Force, $Fi = \pi . d . t . \sigma y(\frac{D}{d} - 0.7)$
5	Sheet metal operations= piercing and drawing and material = mild steel	Stripping force (Piercing) = L x T x 20. L = Length of cut, mm; T = Material Thickness, mm
6	Material = Brass; and Type of press = Double action die	Set the drawing speed = 0.50 m/sec
7	Sheet metal operations= parting off and drawing Aluminium and mild steel	Stripping force (Parting off) =. Stripping force (Drawing) = 30% of Shearing Force X% of Drawing Force.
8	Size of blank in mm ≤ 25.0; and Sheet thickness in mm > 1.0; and Sheet thickness in mm ≤ 2.0; and Die material = Tool steel	Select width of die block in mm = (Strip width + 100.0); and Select length of die block in mm = (Strip length + 110.0)
9	Minimum tolerance required on part in mm ≥ 0.001; and Maximum tolerance required on part in mm ≤ 0.02; and Design feature = hole or slot or oval hole or internal contour cut	Required operation = piercing
10	Inner shape of part feature is circular; and Diameter of inner part feature in mm > 25; and Diameter of inner part feature in mm ≤ 35; and Sheet thickness in mm > 0.5; and Sheet thickness in mm ≤ 0.8	Select circular plain punch; and Diameter of circular punch in mm = (inner circular hole diameter + clearance between punch and die); and Select length of punch in mm = (inner circular hole diameter in mm + 50) ; and Select upper and lower punch plate thickness in mm = 14.0
11	Minimum tolerance required on part in mm ≥ 0.001; and Maximum tolerance required on part in mm ≤ 0.02; and Design feature = hole or slot or oval hole or	Required operation = piercing

IOP Conf. Series: Materials Science and Engineering **244** (2017) 012022 doi:10.1088/1757-899X/244/1/012022

	internal contour cut	
12	Required sequence of operations = piercing and drawing	First operation = drawing and Second operation = Punching
13	Required operations = piercing and blanking	Upper punch = piercing; and Lower punch = blanking
14	Length of die block in mm > 200.0; and Length of die block ≤ 300.0 mm; and Width of die block in mm > 200.0; and Width of die block ≤ 300.0; and Tolerance on part in mm ≤ 0.005 Direction of feed of strip to the die-set = Parallel	Place die in the four pillar die-set with pillar having diameter of bush in mm = 26.0; and Bolster dimension in mm = 34.0; and Length of die-set in mm = 380.0 Width of die-set in mm = 350.0 height of die-set in mm = 28.0

4. Sample run of the proposed system

The sample run and validation of proposed system has been presented through industrial example components shown in Figure 3. User initially loads the system by using the GUI interface with AutoCAD. The system invites the user to enter part data information such as production quantity, tolerance, sheet thickness, sheet material, shear strength etc. through GUI as shown in Figure 4. The system stores these part data automatically in a data file labelled as PART.DAT. As soon as the user enters all required inputs, the system generates the drawings of die components and die assembly in the form of orthographic views and 3D views on AutoCAD screen which are stored automatically in drawing files. Figure 5a, 5b, 5c shows the 2D orthographic views of die component and die assembly automatically generated by the system modules for the example component and Figure 6a and 6b shows the 3D wire frame and 3D solid model of die component and die assembly for the example component. The system is parametric in nature, therefore user can easily modify these drawings through editing respective data files of die components or using AutoCAD commands directly. The system is tested for similar types of parts of different dimensions taken from sheet metal industries. The 2D and 3D drawings automatically generated by system module for component having more than two sheet metal operations are verified from experienced die designers and found very similar to those prepared by the die designers of stamping industries. The time taken by the proposed system for modelling of die components and die assembly for the example part is very less in comparison to the manual drafting and/or by using some CAD software. Notable features of the proposed system are its low cost and the linking of the die design process with quick modelling of die components and die assembly.

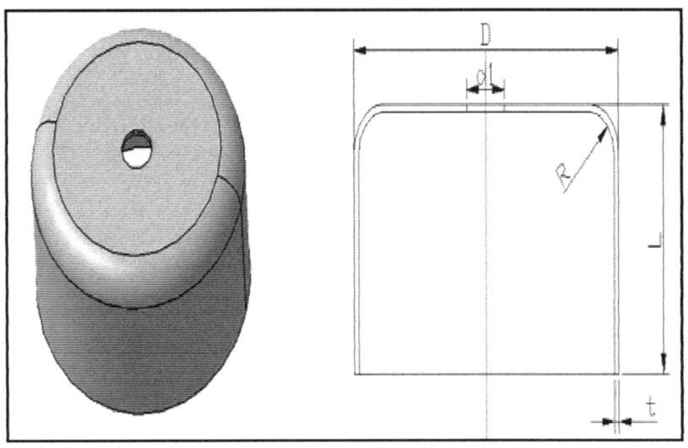

Figure 3. Example component (All dimensions are in mm); Thickness (t) =1mm, Diameter (D) = 99 mm, Pierced hole diameter (d1) = 10 mm, Radius (R) = 8mm, Length (L)=90mm.

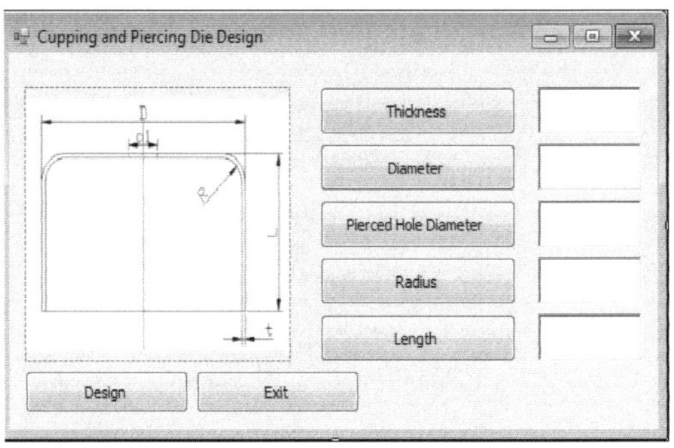

Figure 4. Graphic User Interface (GUI).

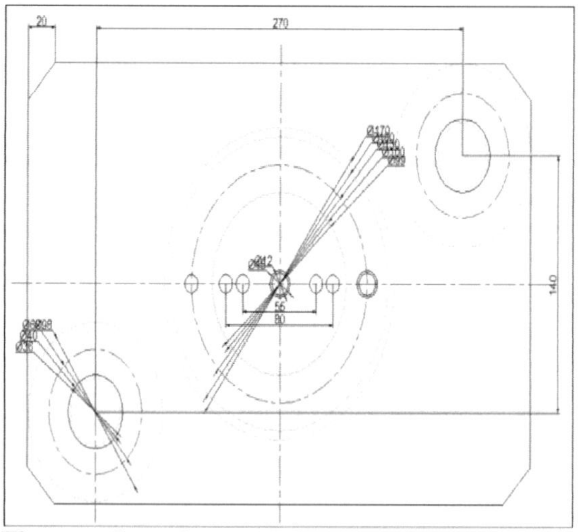

Figure 5a. Output of proposed system forexample component.

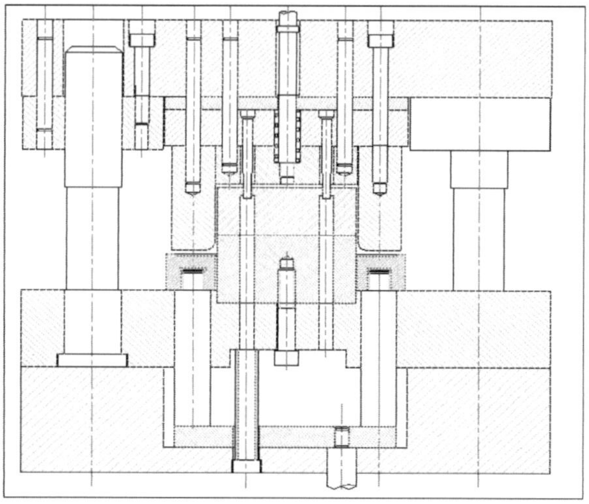

Figure 5b. Output of proposed system for example component.

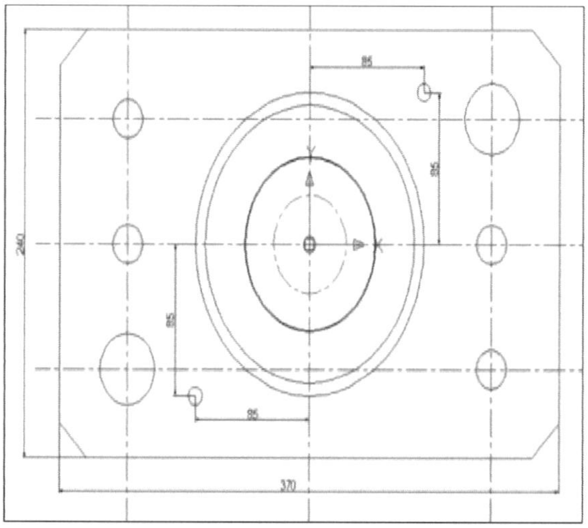

Figure 5c. Output of proposed system for example component.

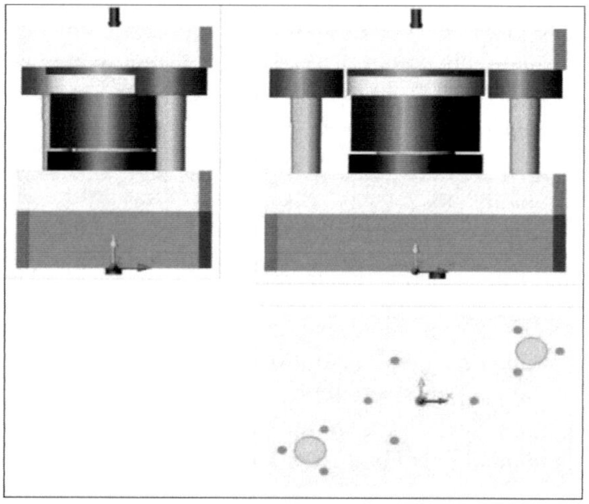

Figure 6a. Output of proposed system for example component in 3D solid mode.

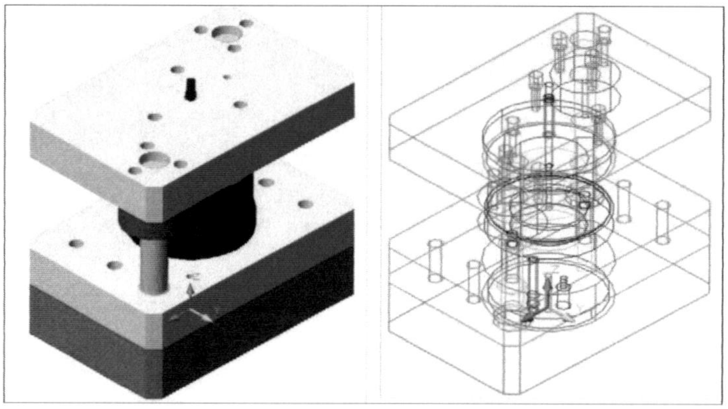

Figure 6b. Output of proposed system for example component in 3D wire frame and 3D solid mode.

5. Conclusion

A knowledge based computer aided system for parametric design of combination die for cupping and punching operations has been developed. The proposed system is coded using Visual basic on AutoCAD platform. A novice engineer who may have little knowledge of die design can do the die design easily using the proposed system. With this system a significant reduction in design and modelling time from days to few minutes has been demonstrated in various industries. The low cost of implementation of proposed system makes it affordable for small and medium scale sheet metal industries. The outputs of the system in the form of dimensional drawings of die components can be used to create formatted data files which are acceptable by the CAM package for manufacturing the die components.

6. References

[1] Wilson F W, Harvey P D and Gump C B (1965) Die Design Handbook (McGraw-Hill Book Co.2e)
[2] Prasad Y K D V and Somasundaram S (1992) Comput. and Control Eng J. 3 185
[3] Choi S H and Wong K W, Int C Mfg Auto, Hong Kong 674
[4] H. S. Ismail, K. Huang, K K B Hon (1992) I-Mech-E, Part B, J EngManuf, 207 117 (1993)
[5] Singh R and Sekhon G S (2003) I-Mech-E, Part B: J Eng Manuf, 217 235
[6] Hussein H M A, Abdeltif L A, Etman M I and Barakat A F (2008) Int C mech design & prod. Egypt,
[7] Cheok B T, Ridong J R, Leow L.F and Nee A Y C (1999) Int C CIM, Tec Singapore 2 1048
[8] Gürün H. and Nalbant M (2005) J. Fac. Eng. Arch. Gazi Univ. 20 155
[9] Kumar S and Singh R (2011) Expert Syst Appl. 38 4482
[10] Choi J C (1988) Korea Sci and EnggFounda, Final Report
[11] Park S B (1997) A study on the Computer-aided system for process planning and CAD/CAM in axisymmetric deep drawing process, Ph. D. Thesis, Pusan National Univ, Korea
[12] Abdel-Magied R K and Wifi A S and Gomaa A H (2004) Conf. on Mech Design and Prod. Cairo 1089
[13] Naranje V. and Kumar S (2014) Expert Syst. Appl.41, 1419 -1431
[14] Zhang Z and Gu W, (1987) Tool and Die Indust. 8 7
[15] Wang F (2005) Composer-CAD User manual
[16] Potočnik D, Dolšak B, Ulbin M (2011) APEM journal Adv. produc. engineer manag. 6 129
[17] Potočnik D, Dolšak B, Ulbin M (2013) J Zhejiang Univ. Sci. A. 14 327
[18] Potočnik D, Dolšak B, Ulbin M (2013) J Braz. Soc. Mech. Sci. Eng., 35 293
[19] Potočnik D, Dolšak B, Ulbin M (2013) Concurr. Eng. Res. Appl. 21 155
[20] Kashid S and Kumar S (2012) J Manuf Eng. 2 168
[21] Kashid S and Kumar S (2012)AJIS 2 168
[22] Kashid S and Kumar S (2013) Int. C Produ. Indust. Eng India 58
[23] Perederli V S (1974) Chem. and Petrol. Eng. 10 171
[24] Ashok-Kumar C N (2012) Appl. Mech. and Mater. 148-149 11
[25] Ashok-Kumar C N, R. Deivanathan (2012)Appl. Mech. and Mater. 234 59
[26] Ashok-Kumar C N., R. Deivanathan (2012) Appl. Mech and Mater. 234 64
[27] Ashok-Kumar C N (2014) IJAER, 9 22 11
[28] Paquin J. R. 1962 Die design fundamentals, Industrial Press

Buildup factor and mechanical properties of high-density cement mixed with crumb rubber and prompt gamma ray study

P Aim-O[1,2], D Wongsawaeng[1], S Tancharakorn[2] and M Sophon[2]

1 Department of Nuclear Engineering, Faculty of Engineering, Chulalongkorn University, Bangkok, Thailand, 10330
2 Synchrotron Light Research Institute, Nakhon Ratchasima, Thailand, 30000
E-mail: doonyapong.w@chula.ac.th

Abstract. High-density cement mixed with crumb rubber has been studied to be a gamma ray and neutron shielding material, especially for photonuclear reactions that may occur from accelerators where both types of radiation exist. The Buildup factors from gamma ray scattering, prompt and secondary gamma ray emissions from neutron capture and mechanical properties were evaluated. For buildup factor studies, two different geometries were used: narrow beam and broad beam. Prompt Gamma Neutron Activation Analysis (PGNAA) was carried out to determine the prompt and secondary gamma ray emissions. The compressive strength of samples was evaluated by using compression testing machine which was central point loading crushing test. The results revealed that addition of crumb rubber increased the buildup factor. Gamma ray spectra following PGNAA revealed no prompt or secondary gamma ray emission. Mechanical testing indicated that the compressive strength of the shielding material decreased with increasing volume percentage of crumb rubber.

1. Introduction

Radiation shielding is based on the principle of radiation attenuation, which is the ability of a shielding material to effectively reduce the radiation's intensity by blocking particles or electromagnetic energy. Some radiation sources emit both neutron and gamma ray from absorption of high-energy gamma ray by atomic nuclei and ejection of protons and neutrons from the nuclei. Therefore, under this situation, neutrons and gamma ray are the main types of nuclear radiation that have to be considered in shielding design [1].

Gamma radiation interacts with the shielding material and become attenuated by photoelectric effect, Compton scattering, and pair production processes. The effectiveness of gamma ray shielding depends on the atomic composition (Z) of the shielding material and the photon energy. Certain gamma ray interaction processes can also result in emission of secondary photons [2].

The transmission of gamma radiation through materials depends on two factors: the attenuation factor and the buildup factor [3]. The attenuation factor accounts for the relaxation length which is the reduction of uncollided photons. The buildup factor accounts for the increased radiation intensity from secondary or scattered radiation and also the number of relaxation lengths [3], which can be computed from the multiplication of linear attenuation coefficient and shielding thickness or μx.

The buildup factor is considerably important for multi-energy gamma-rays with poor geometry for the reason that attenuation coefficients and cross sections of various mediums are not accurate. The

buildup factor is described as the ratio of total value of specified radiation quantity at any point of contribution to the value from radiation transmission at the point without undergoing a collision [4]. Moreover, prompt gamma ray may occur from a neutron interaction process. When a neutron interacts with the target nucleus via a non-elastic collision, the compound nucleus forms in an excited state due to the binding energy of the neutron with the nucleus. The compound nucleus will instantaneously de-excite into a more stable configuration through emission of one or more characteristic prompt gamma rays and secondary gamma rays [5]. This interaction is called the neutron capture or (n, gamma) reaction. Prompt Gamma Neutron Activation Analysis (PGNAA) is a non-destructive method that is useful for simultaneous determination of a trace amount of many elements in samples. This technique measures the prompt gamma rays and secondary gamma ray emission during the neutron interaction.
This work focused on the buildup factor, which is the effect of the scattered and secondary radiation and is a function of the shielding material, the thickness and the energy of the radiation. The shielding material evaluated in the present study was high-density cement mixed with various percentages of crumb rubber [6]. Moreover, the prompt and secondary gamma rays will be considered in this study by the PGNAA method. These two pieces of information will help to completely determine the safety of the shielding material before it can be put into real applications. Moreover, the mechanical properties of the shielding material were evaluated as well.

2. Materials and methods

2.1. Materials and shielding preparation

Natural rubbers are often used in many industries including tyre industries, and are the potential products for neutron shielding. The tyre wastes have been increasing because of the rising demands for automobiles. Crumb rubbers are obtained from tyre wastes by recycling via mechanical process. Crumb rubbers have the special properties to effectively attenuate neutron radiation because they contain a high density of hydrogen and carbon atoms. The shielding material in this work was designed to simultaneously shield against the neutron radiation and gamma radiation, thus, mixing hydrogenous materials, heavy elements and other neutron absorbers help increase the effectiveness of shielding materials [7]. The shielding material in this research composed of Portland cement type I, barite and various percentages of crumb rubber from 5 to 25 vol%, which 5 vol% step increment. All compositions were mixed together and fabricated into a cube shape with dimensions 5 x 5 inches [6].

2.2. Measurement system

2.2.1. Buildup factor measurement and electronic system.

To determine the buildup factor, two geometric arrangements were required, which were narrow beam (good-geometry) and broad beam (poor-geometry). Transmissions of the incident photons were measured, allowing calculation of the attenuation coefficient. For the narrow beam geometry, the gamma radiation was collimated using a cylindrical lead collimator. The thickness of the lead collimator for the source holder was 5 cm with the central hole of 1 cm in diameter and 4.5 cm in height. The diameter of the lead collimator on the detector was 16 cm with the central hole of 3 cm in diameter. The distance between the isotropic source and the detector was 17 cm. The distance between the isotropic source and the shielding material was 12 cm. The broad beam geometry was obtained by simply excluding both collimators. Figures 1 and 2 show the narrow beam and broad beam arrangements, respectively.
The electronic system for radiation detection employed in this study consisted of the following units: NaI(TI) detector (CANBERRA Model 802-2x2, crystal size 2x2 inches), pre-amplifier and multi-channel analyzer (MCA) . Co-60 gamma isotopic source was used to measure 1.1732 MeV and 1.3325 MeV gamma rays emission.

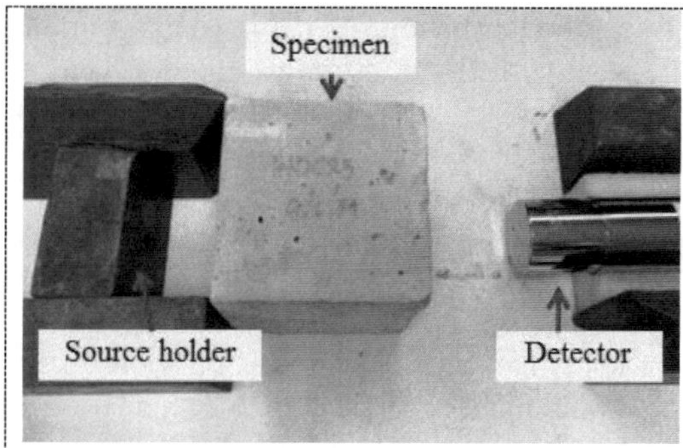

Figure 1. Broad beam arrangement.

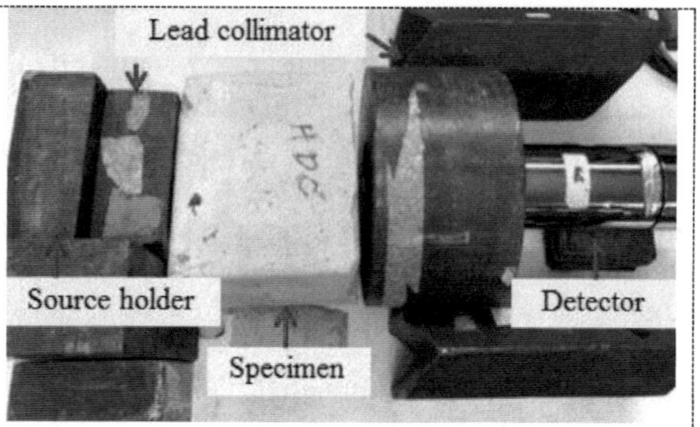

Figure 2. Narrow beam arrangment.

The buildup factor can be calculated by using the total gamma intensity ratio with existence of collimator (good-geometry) or $(I/I_0)b$ divided by the total gamma intensity ratio without existence of collimator (poor-geometry) or $(I/I_0)g$ as follows [8]:

$$B = (I/I_0)b/(I/I_0)g \tag{1}$$

The mean free path (λ), which describes the average distance between two successive photon interactions with a unit of cm, can be calculated from the follow equation [8]:

$$\lambda = 1/\mu \tag{2}$$

Where μ is linear attenuation coefficient in unit cm-1.

2.2.2. Prompt gamma neutron activation analysis.
PGNAA is a measurement technique for non-destructive elemental analysis. A high-purity germanium (HpGe) detector model GC3021 with 30% relative efficiency at 1.1732 and 1.3325 MeV and Am/Be-241 source were used to measure prompt gamma ray emitted from the irradiated specimens. The distance from the neutron source to the specimen was 12 cm. The shielding material was placed 3 cm in front of the detector. The HpGe detector was covered with polyethylene in order to minimize unwanted neutron radiation from getting into the detector and causing damages. This measurement operated at a high voltage of 2,900 V. The system is shown in Figure 3. Each specimen was irradiated for 1 hour using the Am/Be-241 neutron source and the gamma ray spectrum was recorded from 20 to 5,300 keV.

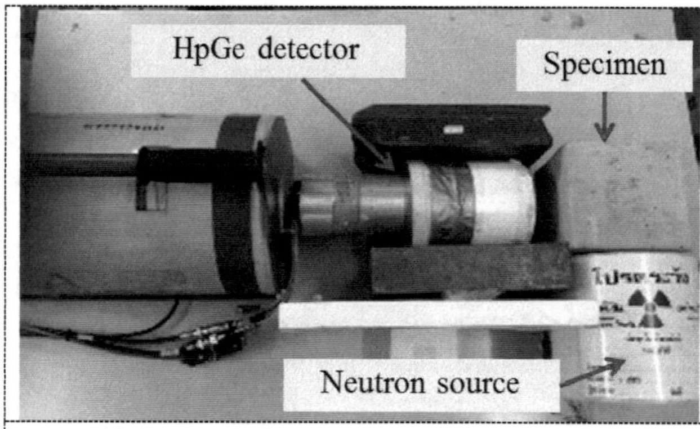

Figure 3. PGNAA experimental setup.

2.3. Compressive strength testing
The compressive strength is the most significant mechanical engineering property of hardening concrete [9]. The compressive strength testing normally requires specimens with a cubical shape [10],

thus, being the reason that specimens in this work were molded into a cubical shape. The compressive strength of samples was evaluated by using a compression testing machine (CTM), which was a central point loading crushing test. Compressive strength testing was conducted after 28 days of curing time.

3. Results and discussions

3.1. Results of mean free path, linear attenuation coefficient and buildup factor

The intensity of broad beam is denoted by Ib and the intensity of narrow beam is denoted by Ig. The results of gamma ray transmissions of narrow beam (good-geometry) and broad beam (poor-geometry) revealed that the ln(I/I0)b and ln(I/I0)g values decreased when the amount of crumb rubbers increased as shown in Table 1. HDC denotes high-density cement; HDCRx denotes high-density cement mixed with x volume percent of crumble rubber.

Table 1. Gamma ray transmissions of narrow beam and broad beam.

Shielding type	I_b	I_g	$\ln(I/I_0)b$	$\ln(I/I_0)g$
HDC	20358.6	21390.7	-1.805	-1.769
HDCR5	22074.6	21427.0	-1.757	-1.768
HDCR10	24161.2	23575.7	-1.635	-1.657
HDCR15	24251.0	23977.0	-1.630	-1.656
HDCR20	22644.0	24508.7	-1.614	-1.634
HDCR25	27386.6	26374.7	-1.509	-1.560

The reason why the $\ln(I/I_0)b$ and $\ln(I/I_0)g$ values decreased with increasing crumble rubber content is because the reduction of barite quantity in the shielding material as a result of addition of crumb rubber. Since barite has the elemental composition known as $BaSO_4$, Ba presents a heavy element with the characteristic properties to attenuate gamma radiation. The high-density cement in this work contained 45.45 w% of barite [6] to function as an effective gamma ray absorber.

Table 2. Mean free path and buildup factor for various type of specimens.

Shielding type	Mean free path	Buildup factor	Standard Deviation
HDC	7.175	0.965	1.023
HDCR5	7.182	1.044	1.091
HDCR10	7.592	1.031	1.158
HDCR15	7.670	1.025	1.497
HDCR20	7.773	1.019	1.159
HDCR25	8.139	1.053	1.218

The mean free path parameter describes the average distance travelled by photons between any consecutive interaction (scattering or absorption). The results revealed that the mean free paths for the case of high density cement (HDC) and high density cement mixed 5 vol% crumb rubber (HDCR5) are lower than the others. Adding more crumb rubber increased the mean free path quite substantially. This behavior can be explained that as the atomic density of rubber is lower than that of HDC and as heavy elements in HDC exhibit large scattering cross-sections for gamma rays compared to H and C, adding more crumb rubber means that gamma rays will be able to travel further before undergoing any interaction. This effectively increased the mean free path of gamma ray observed in the experimental results shown in Figure 4. The effect of crumb rubber addition on linear attenuation coefficient is shown in Figure 5.

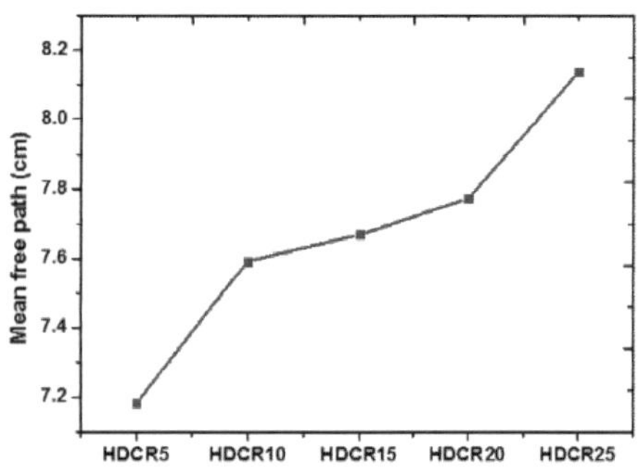

Figure 4. Mean free path of gamma ray after addition of crumb rubber in high density cement.

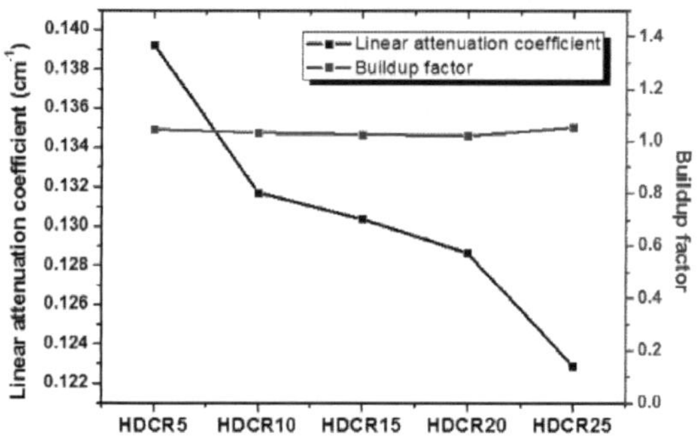

Figure 5. Relationship between linear attenuation coefficient and buildup factors of shielding materials.

The linear attenuation coefficient indicates how effective a shielding material is in attenuating radiation, and is used to calculate the penetration and the energy deposition by photons in shielding materials. The linear attenuation coefficients of high density cement mixed with 5, 10, 15, 20 and 25 vol% crumb rubber were evaluated to be 0.1392, 0.1304, 0.1303, 0.1286 and 0.1228 cm^{-1}, respectively, indicating that the value decreased with increasing crumb rubber addition. This result is expected because as the crumb rubber percentage increased, it effectively lowers the overall Z value and the density of the shielding, resulting in less gamma ray attenuation. In fact, the linear attenuation coefficient must have the opposite trend compared to the mean free path with increasing crumb rubber percentage, and this is clearly observed in results presented Figures 4 and 5.

The results of the buildup factor experiment revealed that the buildup factors of high density cement mixed with 5, 10, 15, 20 and 25 vol% crumb rubber were 1.044, 1.030, 1.025, 1.019 and 1.053, respectively. Normally, the value of the buildup factor should be unity for the case of a collimated beam with monoenergetic photon and should be greater than 1 for other cases [11], for the reason that some scattered gamma ray occurred from shielding materials. The reduction of the buildup factor with increasing crumb rubber content up to 20 vol% is a highly desirable feature because it indicates that addition of crumb rubber helped reduce secondary and scattered gamma ray emissions. Moreover, all of the evaluated buildup factor values were very close to unity, indicating that secondary or scattered gamma radiation were effectively not present.

3.2. Prompt gamma neutron activation analysis results

Figure 6 illustrates the gamma ray spectra obtained from the PGNAA studies. Because the spectra of all shielding designs were very similar, only those for HDCR10 and HDCR20 were shown in the Figure for simplicity. The spectra were also not different when compared with Co-60 background spectrum.

It can be concluded that there was no detectable prompt or secondary gamma ray emission, which is highly desirable. This was a highly desirable feature. However, this does not mean that there was no neutron capture interaction occurring in the shielding material. It means that all the heavy and light elements in the shielding effectively shielded all of the generated prompt and secondary gamma ray. Heavy elemental composites in specimens, mainly being barite, help increase the inelastic scattering interaction, while light elemental composites such as hydrogen and carbon atoms enhance the elastic scattering interaction.

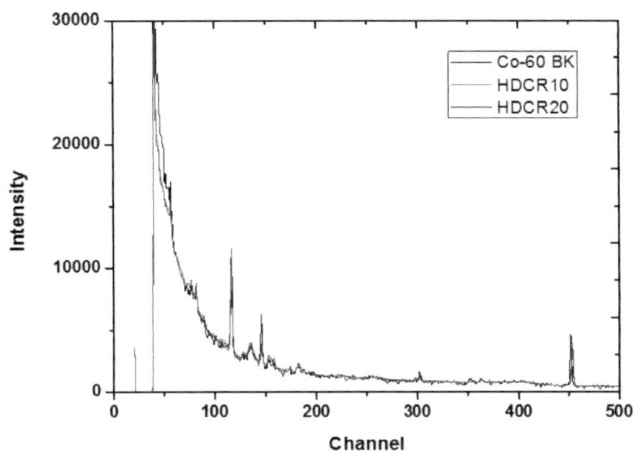

Figure 6. Comparative prompt gamma ray spectra for different types of shielding materials.

3.3. Compressive strength results

Compressive strength testing was conducted to determine the mechanical property of specimens. The results in Table 3 clearly indicated that more crumb rubber addition resulted in less compressive strength. This result is expected because crumb rubber is mechanically much weaker and less dense than the HDC matrix.

Table 3. The 28 days Compressive Strength on different type of shielding materials.

Shielding type	Crushing load (kN)			Compressive strength (N/mm^2)			Avg Compressive strength (N/mm^2)
	1	2	3	1	2	3	
HDC	199.8	297.9	245.6	12.063	18.312	14.725	15.03
HDCR5	139.8	151.6	158.1	9.9770	9.7430	9.6790	9.80
HDCR10	142.5	169.7	174.9	8.3610	10.087	10.610	9.69
HDCR15	183.0	124.6	166.7	11.022	7.4120	9.9710	9.47
HDCR20	129.4	165.5	144.0	7.8980	10.001	8.6930	8.86
HDCR25	104.5	114.7	111.0	6.4330	6.9730	6.7910	6.73

Factors that can contribute to increase the compressive strength of concrete are mix ratio, size and texture of aggregate, method of compaction and curing periods [10]. Thus, the compressive strength of the shielding with addition of crumb rubber could be enhanced by appropriately adjusting the above-mentioned factors. However, although the shielding becomes mechanically weaker with addition of crumb rubber, the shielding should not be designed and used to physically withstand any load other than the weight of the shielding itself. Several methods can be easily employed to mechanically support the shielding material such as encasement in a stronger material.

4. Conclusions

In this work, the buildup factors as well as scattered and secondary gamma radiation of high-density cement mixed with various volume percentages of crumb rubber were evaluated. The results found that the buildup factors for HDCR5 and HDCR25 were 1.044 and 1.053, respectively, which were very close to unity, indicating that secondary or scattered gamma radiation were effectively not present. Gamma ray spectra from PGNAA analysis revealed no prompt or secondary gamma ray emission from neutron interactions with shielding materials. This was a highly desirable feature. The results of compressive strength testing indicated that the compressive strength of the shielding materials decreased with increasing volume percentage of crumb rubber, with the compressive strengths of HDCR5 and HDCR25 of 9.800 and 6.733 N/mm^2, respectively. Several methods can be easily employed to mechanically support the shielding material such as encasement in a stronger material. In conclusion, the safety of the high-density cement mixed with various percentages of crumb rubber was confirmed because the material exhibited several highly desirable features to shield against neutron and gamma radiation simultaneously. This shielding material can be utilized in real applications to ensure radiation safety of personnel. In order to have a complete assessment of the performance of the shielding materials, the usable lifetime of the shielding with respect to gamma ray dose and neutron fluence should be evaluated as well.

5. References

[1] "Photonuclear Reaction", *TheFreeDictionary.com*, 2017. [Online]. Available: http://encyclopedia2.thefreedictionary.com/Photonuclear+Reaction. [Accessed: 30- Jun- 2017]

[2] Jubair, S.I., *Calculation of Buildup Factors for Ceramic Materials.* Iraqi Journal of Applied Physics, 2011. **7** (No.1): p. 23-26

[3] 2017. [Online]. Available: http://web.mit.edu/22.09/lab_writeups/shield.pdf. [Accessed: 30- Jun- 2017]

[4] V. Singh and N. Badiger, "A Comprehensive Study on Gamma-Ray Exposure Build-Up Factors and Fast Neutron Removal Cross Sections of Fly-Ash Bricks", *Journal of Ceramics*, 2013, vol. **2013**: pp. 1-13

[5] "Prompt Gamma-Ray Activation Analysis (PGAA)", *NIST*, 2017. [Online]. Available: https://www.nist.gov/laboratories/tools-instruments/prompt-gamma-ray-activation-analysis-pgaa. [Accessed: 30- Jun- 2017]

[6] P. Aim-O, D. Wongsawaeng, P. Phruksarojanakun and S. Tancharakorn, "Monte carlo simulation of innovative neutron and photon shielding material composing of high density concrete, waste rubber, lead and boron carbide", Journal of Physics: Conference Series, vol. 860, p. 012043, 2017

[7] Hausi Hu, Q.W., Juan Qin, Yuelei Wu, Tiankui Zhang, Zhongsheng Xie, Xinbiao Jiang, Guoguang Zhang, Hu Xu, Xiangyang Zheng, Jing Zhang, Wenhao Liu, Zhenghong Li, Boping Zhang, Linbo Li, Zhaohui Song, Xiaoping Ouyang, Jun Zhu, Yaolin Zhao, Xiaoqin Mi, Zhengping Dong, Cheng Li, Zhenyu Jiang, and Yuanping Zhan, *Study on Composite Material for Shielding Mixed Neutron and Gamma-Rays.* IEEE Transaction On Nuclear Science, 2008. Vol. **55**: p. 2376-2384

[8] M.J.R._Aldhuhaibat, , M.K. Alfakhar, and M.S. Amana,_*NUMERICAL BUILDUP FACTOR CALCULATION OF GAMMA RAYS FOR SINGLE, DUAL, AND MULTI-LAYERS SHIELDS USING LEAD AND ALUMINUM.* International Journal of Recent_Scientific Research, 2015, Vol. **6** (Issue 7): p. pp.5184-5189

[9] X.-Y. Wang and K.-B. Park, *Analysis of compressive strength development of concrete containinghigh volume fly ash.* Construction and Building Materials, 2015, vol. **98**: p. 810–819

[10] B. AHMED and M. RAHMAN, *Assessment of Compressive Strength of Concrete Based On Combination of Different Sizes of Aggregate.* International Journal of Advanced Structures and Geotechnical Engineering, 2015, vol. **04** (No. 03): pp. 148-152

[11] 2017. [Online]. Available: http://mragheb.com/NPRE%20402%20ME%20405%20Nuclear%20Power%20Engineering/Attenuation%20of%20Gamma%20Radiation.pdf. [Accessed: 30- Jun- 2017]

Acknowledgement

The authors would like to thank the Royal Thai government and Synchrotron Light Research Institute (Public Organization) for providing the scholarship to perform this research and Nuclear Engineering Department at Chulalongkorn University for supporting the nuclear instruments.

Enhancement of mechanical properties of porous aluminum by silica sand particles

S Khamsuk A Joosawat N Panomtang and K Wongtimnoi

Department of Industrial Engineering, Faculty of Engineering, Burapha University, Thailand
E-mail: sunisak@eng.buu.ac.th

Abstract. Porous aluminum was fabricated using space holder process. The spherical carbamide powder and silica sand were used as space holder and reinforcement materials, respectively. Aluminum powders were mixed with five different silica sand contents and were then mixed with 10 wt.% carbamide powder. The mixed powders were cold compression at 400 MPa and sintered at 550 °C for 5 hours. Microstructure of specimen was characterized by scanning electron microscope. Hardness and compressive strength were also investigated in this study. It was found that microstructure and mechanical properties of porous aluminum strongly depended on silica sand contents. From SEM observation, it is clearly seen that addition of silica sand led to increasing pore in the porous aluminum. Mechanical properties of porous aluminum can be enhanced by silica sand particle reinforced.

1. Introduction

During last two decades, porous aluminum has become an attractive topic in materials science research and also in the industry, especially in the automotive industry [1]. These are owing to their unique mechanical and physical properties such as high strength to weight ratio, light weight, high energy absorption and sound absorption [2]. Although, it has the excellent mechanical properties as mentioned earlier its strength is not appropriate for some engineering applications. Therefore, it is necessary to improve its mechanical property for fulfilling the demands. One promising method for improvement of porous aluminum strength is reinforced with particles. Several elements have been successfully used for enhancing of dense aluminum mechanical property i.e. copper, magnesium, silicon, scandium, alumina, silica sand and so on[3-9]. It is well known that the competition in an industry is relatively high, especially in terms of the product price. Thus, to sustain their business, all the companies need to reduce the manufacturing cost. From this point of view, silica sand has become a promising reinforcement material because it is extremely cheap compared with the other reinforcement particles. On the other hand, many researchers have found that the hardness of dense aluminum increased more than one time when silica sand was added [8-9]. However, it should be noted here that it has rarely been used in porous aluminum which has relatively less grain boundary.

Recently, porous aluminum can be produced by various processes such as melt gas injection, melts with blowing agents, casting around space holder material, powder compact process. It has found that those processes have limitation on the controlling the pore shape and distribution [10]. Therefore, space holder method has become a promising method because the morphology of pore can be controlled more precisely [11-13]. This fabrication method involves four steps, e.g. mixing based material with space holder particles, compaction, remove space holder particles and sintering. Carbamide powder is one of the most widely used space holder materials because it is cheap and easy to purchase. Therefore, carbamide powder is selected as space holder material in this study.

IOP Conf. Series: Materials Science and Engineering **244** (2017) 012024 doi:10.1088/1757-899X/244/1/012024

In the present work aimed to improve mechanical properties of porous aluminum by reinforced with silica sand particles. Microstructure and mechanical properties of porous aluminum will be systematically studied.

2. Experimental procedure

High purity aluminum powder with a purity of 99.9 % and mean particle size of 44 μm supplied by Supbon Co., Ltd. was used as a starting material. The aluminum powder used in this study has an irregular shape as shown in Figure 1 (a). Silica sand powder from Trad province, Thailand was used as a reinforcement material. The received silica sand was firstly cleaned by water to remove impurity, dried, mechanical grounded for refining the particle size and then sieved with a 300 μm sieve, respectively. Spherical carbamide powder with an average particle size of 2.4 mm was used as space holder material. Morphologies of silica sand and carbamide powders were shown in Figure 1 (b) and (c), respectively.

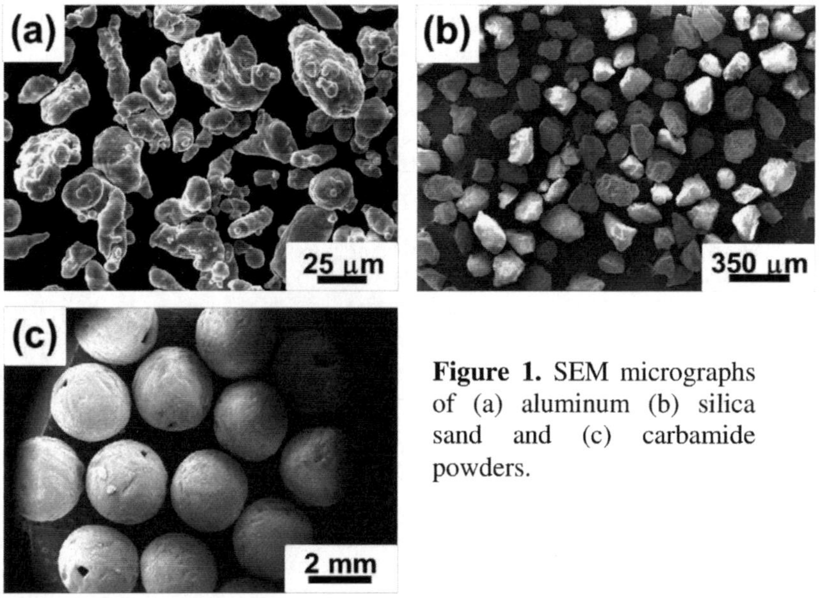

Figure 1. SEM micrographs of (a) aluminum (b) silica sand and (c) carbamide powders.

In order to produce porous aluminum, the aluminum powder was firstly mixed with 5, 10, 15 and 20 wt.% silica sand powder by mixing machine at a constant mixing speed of 100 rpm for 4 hrs. Then the mixed silica sand/aluminum composite powders were continuously mixed with 10 wt.% carbamide powder by mixing machine under the same mixing speed for one more hour. In this study, the high purity aluminum powder was also mixed with 10 wt.% carbamide powder under the same mixing conditions for comparison. After that, all mixed powders were compacted in a cylinder die with a constant pressure of 400 Mpa at room temperature. Subsequently, green compacted specimens were removed space holder materials by placed in ultrasonic water bath at 80 °C for 1.5 hrs. Afterward, the green porous were sintered at 550 °C for 5 hrs with a heating rate of 10 °C/min. After sintered, the specimens were furnace cooled to room temperature. The microstructure of porous aluminum was characterized by LEO 1450 VP scanning electron microscopy. Specimens for microstructure analysis were prepared by mechanical polishing using 600, 800, 1200 and 2000 grit SiC papers, followed by ached in a solution of 6 vol.% nitric acid (HNO_3), 12 vol.% hydrochloric acid (HCl), 1 vol.% hydrofluoric acid (HF) and 81 vol.% distilled water for 15 seconds. Density and porosity of porous aluminum were calculated using the classical equation 1 and 2, respectively.

$$\rho_s = \frac{W_d}{W_w - W_s} \times \rho_{water} \tag{1}$$

$$Porosity = (1 - \frac{\rho_s}{\rho_T}) \times 100 \qquad (2)$$

$$\rho_T = \rho_{Al} V_{Al} + \rho_{sand} V_{sand} \qquad (3)$$

Where Wd is dry weight of specimen, Ww is weight of suspended spacemen in the water, Ws is weight of saturated specimen, ρ_s, ρ_{Al} and ρ_{sand} are density of specimen, aluminum and silica sand (density of aluminum and silica sand are 2.7 and 2.65 g/cm^3, respectively), V_{Al} and V_{sand} are volume fraction of aluminum and silica sand, respectively.

Hardness test was measured using microhardness tester model Innova test Europe BV equipped with 10 mm Brinell indenter at 100 kgf of a load. Hardness test was conducted randomly at five different points per one specimen. The average hardness value was computed from three values exclude the minimum and maximum values. The compressive test was carried out by a 10 kN load universal testing machine with a cross head speed of 1 mm/min. The specimens used for the compression test were prepared according to ASTM E9-89a that is 13 mm in diameter and 25 mm in height.

3. Results and discussion

3.1. Density and porosity

Density and porosity of porous aluminum plotted as a function of silica sand content is shown in Figure 2. It can be seen from this figure that the density gradually decreased with increasing amount of silica sand. This is attributed to silica sand has lower density than aluminum. Therefore, it causes a decreasing in porous aluminum density when increases silica sand powder content. It was also found that porosity of porous aluminum increased with increasing silica sand content. Similar observations have been found in silica sand/aluminum composite fabricated by powder metallurgy route [7-9]. This may cause by silica sand is incredibly harder than aluminum. Thus, silica sand probably not plastically deform like aluminum during the compaction process. Moreover, it is also had a different thermal expandsion and shrinkage with aluminum. So, that may cause pore between aluminum and silica sand interface.

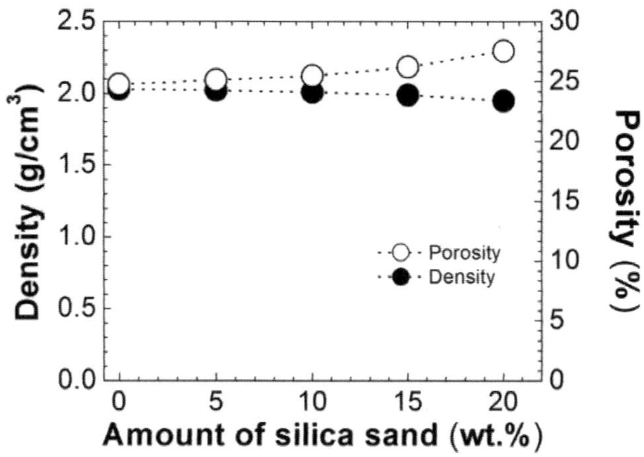

Figure 2. Density and porosity of porous aluminum versus amount of silica sand.

3.2. Microstructure evolution

Microstructure of porous aluminum at various silica sand contents are shown in Figure 3. It can be seen clearly from Figure 3 that the specimens consisted of rounded pores with an average pore size of 2.4 mm, which is same as the initial characteristic of the carbamide particles. This is attributed to the decomposition of carbamide particles during the space holder removal process caused the pores in specimen. Thus, the morphology of pores was similar with space holder materials. These results indicated that the shape and size of pores in porous metallic materials fabricated by this method can be controlled by morphology of space holder materials.

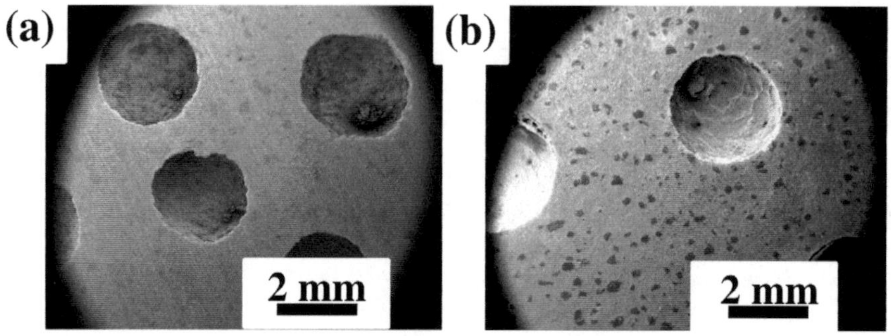

Figure 3. *SEM* micrograph *of porous aluminum-silica sand (a) 5 %wt. and (b) 10 %wt. silica sand (8x).*

In order to investigate the influence of silica sand powder on microstructure evolution in porous aluminum, the microstructure observation was performed with a higher magnification at the cell wall regions. SEM micrographs of cell wall porous aluminum at different silica sand content ranging from 0 wt.% to 20 wt.% are shown in Figure 4 (a) – (e), respectively. Porous aluminum without silica sand has high density (Figure 4 (a)). Microstructure of porous aluminum with addition of 5 wt.% in Figure (4 (b)) showed that the specimen consisted of uniformly second phase distribution. Second phase particles had well bonded with aluminum matrix. With increasing silica sand to 10 wt.%, voices between silica sand and aluminum bonding surface was observed (Figure 4 (c)). Amount of voices at the interface second phase particle regions increased with an increasing silica sand content (Figure 4(d) and (e)). These results obviously indicated that increasing amount of silica sand is lead to increase amount of pore in aluminum. This results is well consisted with the porosity measurement results (Figure 2).

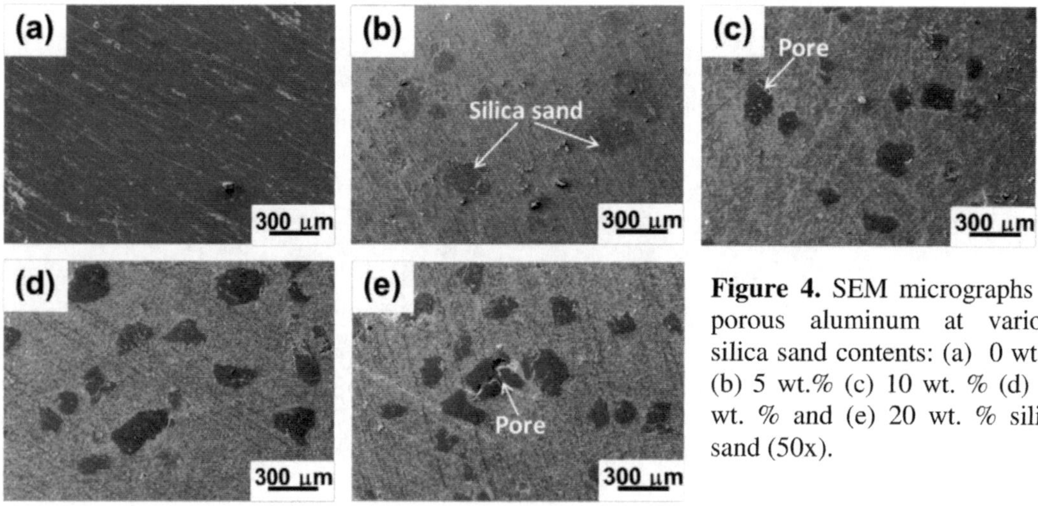

Figure 4. SEM micrographs of porous aluminum at various silica sand contents: (a) 0 wt.% (b) 5 wt.% (c) 10 wt. % (d) 15 wt. % and (e) 20 wt. % silica sand (50x).

3.3. Hardness

The hardness of the porous aluminum is plotted as a function of silica sand content in Figure 5. The hardness of porous aluminum enhanced from 15.12 HB to 17.35 HB when added 5 wt. % silica sand. Hardness continuously increased with increasing amount of silica sand up to 15 wt.%. Future increasing silica sand particles, the hardness decreased to 20 HB. It can be seen from this results that the hardness of porous aluminum is strongly depended on the silica sand content. Hardness of porous aluminum enhanced 75 % when added 15 wt.% of silica sand.

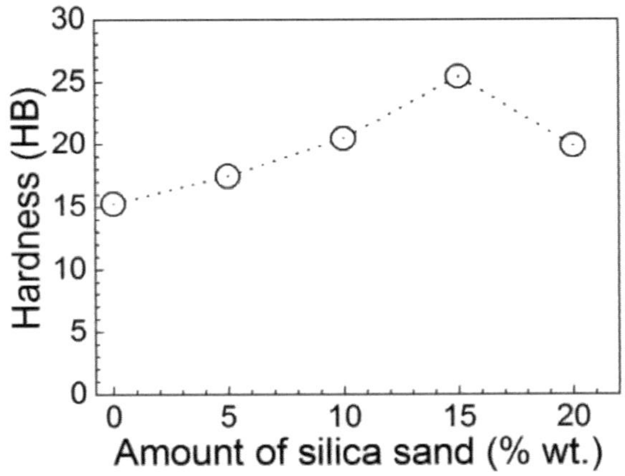

Figure 5. Hardness of porous aluminum versus amount of silica sand.

3.4. Compressive test

The compressive stress-strain of porous aluminum at five different silica sand contents is shown in Figure 6. The compressive stress-strain curve of porous aluminum exhibited three different stages that is linear elastic deformation, plateau deformation and densification deformation stage, which is usually observed in the porous metallic materials [11,12,14]. Compressive strength of porous aluminum showed the similar trend with hardness test results. That is the strength increase with increasing silica sand content up to 15 wt.% and then slightly decreased with future increasing of silica sand particle. The increasing of strength in porous aluminum when silica sand was added is caused from the present of second phases inhibited dislocation motion and harden material. These results indicated that the mechanical property of porous aluminum can be enhanced by addition of silica sand particles.

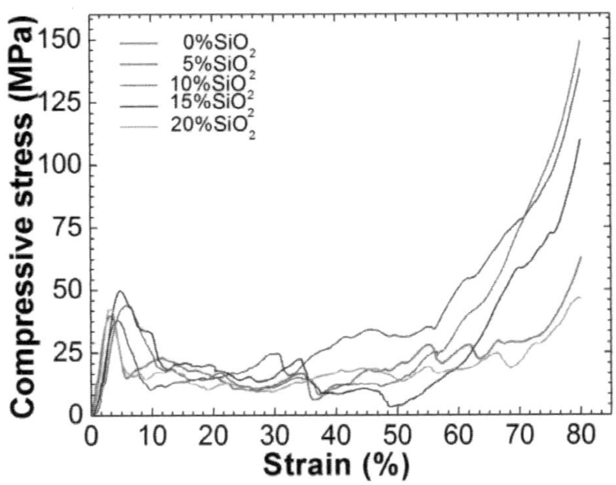

Figure 6. Compressive stress-strain curves of porous aluminum at different amount of silica sand.

4. Conclusions

Porous aluminum was synthesized by space holder method by using spherical carbamide powder as space holder material. Silica sand particles were purposely added to enhance mechanical properties of porous aluminum. The influence of silica sand content on microstructure evolution and mechanical properties of porous aluminum were investigated. The main results obtained from this study are summarized as follows:

1. The density of porous aluminum decreased and porosity increased with increasing of silica sand content.
2. The debonding of silica sand particle with aluminum matrix was observed when added silica sand above 5 wt.%.
3. Hardness and strength of specimens were increased with increased amount of silica sand up to 15 wt.% silica sand and then decreased with future increasing of volume of silica sand. Mechanical properties of porous aluminum enhanced 75 % when added 15 wt.% silica sand particles.

5. References

[1] Baumeister J, Banhart J and Weber M 1997 Mater. Des. Vol. 18, p.217-220
[2] Jeon Y, Kang C and Lee S 2009 J. Mater. Process. Technol. Vol.209(1), p.435–444
[3] Rana R, Purohit R and Das S 2012 IJSRP. Vol. 2
[4] Khamsuk S, Park N, Terada D and Tsuji N 2016 Mater. Sci. Forum. Vol. 872, p.33-37
[5] Sun Y, Song M and He Y 2010 Rare Metals. Vol. 5 p.451-455
[6] Verma J, Kumar A, Chandrakar R and Kumar R 2012 JMMCE Vol.11, p.1126-1131
[7] Rizkalla H and Abdulwahed A 1996 J. Mater. Process. Technol. Vol. 56, p.398-403
[8] Zuhailawati H, Samayamutthirian P and MohdHaizu C 2007 IJPS. Vol. 18(1), p.47–55
[9] Ahmed T and Mamat O 2011 Solid State Science and Technology Vol. 19, p.138-149
[10] Banhart J 2001 Pro. Mater. Sci. Vol.46, p.559–632
[11] Hussain Z and Suffin N 2011 J. Eng. Sci. Vol. 7, p.37–49
[12] Jiang B, Zhao N, Shi C, Du X, Li J and Man H 2005 Mate Lett. Vol. 59, p.3333 – 3336
[13] Surace R, Filippis L, Ludovico A and Boghetich G 2009 Mater. Des. Vol. 30 p. 1878–1885
[14] Bafti H and Habibolahzadeh A 2013 Mater. Des. Vol. 52, p.404-411

Acknowledgements
The authors gratefully acknowledge Faculty of Engineering, Burapha university through the grant number VJP. 1/2560 for financial support.

IOP Conf. Series: Materials Science and Engineering **244** (2017) 012025 doi:10.1088/1757-899X/244/1/012025

Study on service performance of 880 MPa-grade and 980 MPa-grade rail steels

Zhu M[1,2], Xu G [1,*], Zhou J H[2], Wang R M[2] and Gan X L[1]
1 School of Materials and Metallurgy, Wuhan University of Science and Technology, Wuhan, 430081, China;
2 Research and Development Center, WISCO, Wuhan 430080, China

Abstract: With the rapid development of economy in China, the requirement for railway passenger and cargo transportation becomes higher and higher. Increasing speed and developing heavy-haul transportation can effectively improve the transportation capacity of railway. The requirement for rail steels with higher ability of abrasion resistance becomes urgent. Two kinds of rails, i.e., 880 MPa-grade and 980 MPa-grade rail steels, were laid on cargo line with 500 m short radius curve. The service performances of the tested steels were continuously tracked and analyzed during operation of total 22.5 million tons loads. Macro morphology, light band width, surface hardness and profile of rails were investigated. The results show that the performance of 980 MPa-grade rail steel is superior to that of 880 MPa-grade rail steel at same circumstance. Therefore, 980 MPa-grade rail should be selected in the cargo line with small radius curve.

1. Introduction

With the continuous and rapid development of China's economy, the demand for passenger and cargo flow is increasing, and the demand for railway transportation is becoming higher and higher [1-2]. At the end of 2016, China's railway mileage has reached 124,000 kilometers. Railway transportation plays an important role in China's passenger and freight transportation, which directly relates to the development of the national economy.

China's passenger and freight transportation speed up in large areas, and high-speed heavy haul railway develops rapidly, which makes rail service life seriously reduce [3-4]. At the same time, the difficulty of maintenance of rail damage due to the increased traffic density is constantly increasing, resulting in more potential traffic safety hazards [5]. The situation of rail damage is increasingly grim [6]. Rail is an important part of the orbital structure in railway transportation, which supports and guides the train [7]. Rail abrasion [8-9] is the main type of damage in the curve section of railway, and rail abrasion has become the main factor in the damage of curve rail. In the process of using rails, the earliest defects occur in curved segments [10], especially in the small curve radius rail line ($r \leq 800$ m). The wheel and rail angle contact and the rail force is complex. Under the wheel-rail alternating stress cycle, the rail is prone to fish scale injury, severe abrasion, and so on [11]. In order to improve the service life of rail and reduce the safety hazard, some investigations have been conducted on rail wear. Sato Yoshihiko [12] studied the service situation of the rails and proposed the method of service life extension of the rail. Krylov [13] introduced the maintenance operation of sinkansen in Japan structure. Bo [14] claimed that the main factors that produce curved rail side grinding are the wheel's impact angle on rails and the two-point contact of the wheels. Wang et al. [15] carried out a long-term investigation and analyzed the differences between the high speed and heavy-haul railway damage. So far, the wear situation of 880 MPa-grade and 980 MPa-grade on the rail line of small curve radius (r ≤ 800 m) is rarely reported. At present, the 880 MPa-grade and 980 MPa-grade rails are mostly used in China, which accounts for 90% of the annual total amount of rail. In order to study the performance of rail service of different strength levels, especial in small curve radius rail line, prolong the service life of rails and promote the rational use of rails, improve the comprehensive use efficiency, 880 MPa-grade and 980 MPa-grade rails are selected in the present study to lay on the small curve radius railway respectively. The service performance of two different rails were observed and analyzed during transportation of total 22.5 million tons of loads. The abrasions of rails in small curve radius railway were investigated.

2. Experimental methods

On a railroad line in Wuhan, China, three rails of 60 kg/m with a length of 12.5 meters and different strength of 880 MPa (U71Mn) and 980 MPa (U75V, U68CuCr) rails were continuously inserted on the 500 m curve radius railway line. Rails were bonded by the splints. The test line is used for cargo transportation. The axle weight of the trains is 23 t and the annual total weight is 13 Mt. The properties of test rail are shown in table 1. Firstly, the contact and force between wheel and rail in the line segment and the curve segment were simulated by finite element software. Actual test time is 1.5 years. Macro morphology, light band width, surface hardness and profile of rails were tracked and analyzed for three times corresponding to three different conditions of total weight of 40000 tons, 0.92 million tons and 22.5 million tons (corresponding to the running time of 1 day, 1 month, 1.5 years, respectively). The rail performances of different strength levels under the same test condition were compared and analyzed.

Table 1. The performance of experimental rail.

Steel grade	Rm (MPa)	A (%)	HBW10 / 3000(HB)
U71Mn	940	13	268
U75V	1040	12	301
U68CuCr	1045	12	302

3. Experiment results and discussions

3.1. Analysis of force bearing point on rails

By means of digital simulation, the contact and force between wheel and rail in the straight and curved segments are simulated. The results shows that when the train is traveling in a straight segment, the contact and force mode between rail upper and lower stocks and wheel are the same. The wheel and the rail top contact, with the feature of the minimal wheel-rail contact stress and the very small lateral creep force, which is one of the ideal states of wheel-rail contact. However, when the train is in curve segment, in the contact area between upper rail gauge angle and the wheel rim root, there is relative sliding and rotation between the wheel and rail contact surfaces, which can usually produce a large wheel margin force and lateral creep, resulting in longitudinal and lateral forces. Therefore, the probability of wear and damage of rails on the curve segment is the largest [16-17].

3.2. Macroscopic morphology

The rails of 880 MPa and 980 MPa show different macroscopic morphology after service for 1.5 years with load. There is no obvious change in rail bottom and rail waist for three rails. But the contact surfaces between rail head and wheel change obviously. The 880 MPa rail mainly manifests lateral grinding and vertical grinding of rail. The rail bottom distributes obvious steel scraps due to the abrasion of wheel and rail (see Figure 1), while there is no obvious fish scale, peeling off the block and so on the surface of rail head. In the early stage of service for 980 MPa rail, there is a small amount of fish scales which display 45° angle distribution to rail axial at work side. With the increase of the total weight through the rails, the number of fish scales increases and more phenomena of peeling off occur, as shown in Figure 2.

With the increase of total weight, the rail is subjected to the vertical pressure of train, and varying degrees of width spread appear on rails. When the total amount is 0.92 million tons, the thickness of width spread for U71Mn rail reaches to 1.8 mm, whereas the thickness of width spread for U75V and U68CuCr rails is 1 mm. With the further increase of the total weight, the width spreads are gradually worn out and the macroscopic topography of the rail mainly manifests lateral grinding and vertical grinding.

Figure 1. The macro morphology of 880 MPa-grade U71Mn rail.

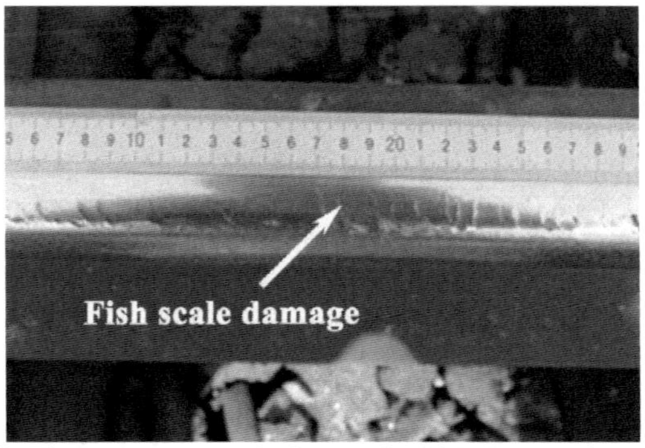

Figure 2. The macro morphology of 980 MPa-grade U75V rail.

3.3. Hardness of the rail working face

The rail surface will produce plastic deformation due to the force because the rail and the wheel repeatedly contact. The proliferation and entanglement of dislocation produce a lot of sub-grains, leading to work hardening. With the increase of total weight, the hardness of rail head surface gradually increases. The higher the hardness of rail surface is, the faster the plasticity of rail surface decreases. The rail is more prone to wear and tear off the block in the process. It can be seen from Figure 3, the initial hardness of 80 MPa-grade (U71Mn) rail is low and the surface hardness increases faster than 980 MPa-grade rails (U75V and U68CuCr) with the increase of the total weight. The wear is easier to happen for 880 MPa-grade rail (U71Mn) due to the fast decrease of plasticity caused by more serious work hardening.

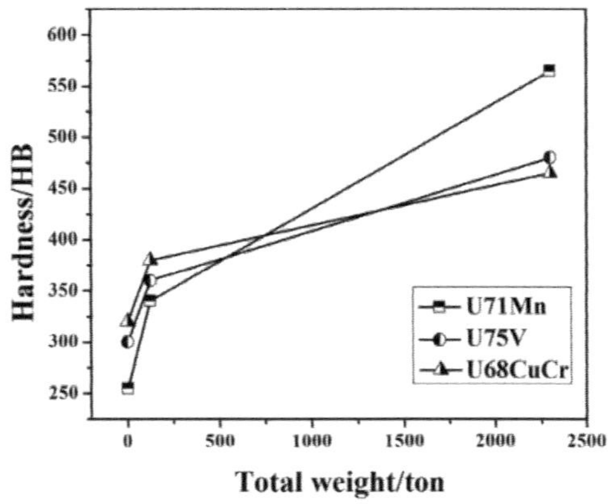

Figure 3. The hardness of rail surface.

3.4. Change of light band width

The width of light band is an indicator to measure the ability of pressure resistance of rails. The wider the light band is for the same original profile after passing the same total weight, the lower ability of capacity of pressure resistance of rails is and the more serious the rail deformation is. Under the load pressure, the greater the rail deformation is, the mire contact area of the vehicle is and the wider the light band is. Figure 4 shows the relationship between the width of light band and the total weight. It is observed that with the increase of total weight, the light bands for 980 MPa-grade (U75V and U68CuCr) and the 880 MPa-grade (U71Mn) rails significantly increase in the early stages and subsequently slowly increase. In the early stage of wear, the increase of bright band for 880 MPa-grade (U71Mn) rail is more obvious than the 980 MPa-grade (U75V and U68CuCr) rails. The reason is that the hardness of former surface is lower than the latter. The contact area between wheel and rail is larger under the load rail and it is more prone to produce light belts. It can be seen from the hardness of rail surface in Figure 3, although the final hardness of 880 MPa-grade rails is greater than that of 980 MPa-grade rails, the former light band is close to the latter total bright belt in the early stage of wear (920,000 tons, corresponds to a month). When the total weight reaches 22.5 million tons, the light belt of 880 MPa-grade rail is 8 mm wider than that of 980 MPa-grade rails.

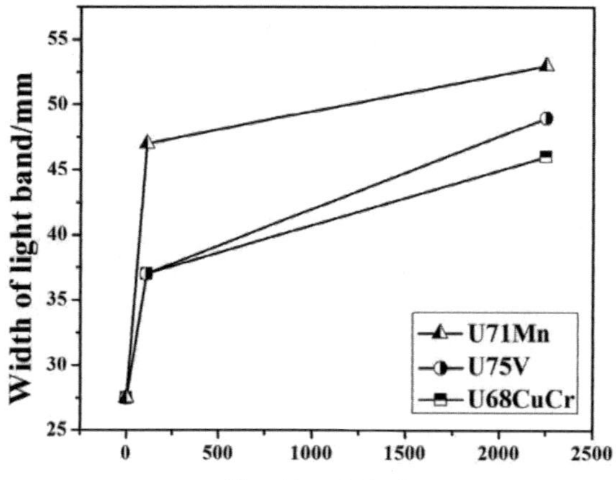

Figure 4. The total weight vs light band width.

3.5. Rail profile

The initial profile of test rails is the standard profile of 60 kg/m rails in TB/T 2344, China rail national standard. Deformation and abrasion happen on the rail because of the service under laod. The rail profile slowly changes due to continuous abrasion of wheel and rail. The performance of rail use can be directly reflected by measuring the profile of rail head.

The measurement position of vertical abrasion of rail is offset 10 mm to rail working side form rail head centerline. The measurement position of lateral abrasion of rail is offset 16 mm form rail head. The rail profiles were measured by rail profile instrument and amount of the lateral grinding and vertical grinding was calculated by comparing real rail profiles with the standard contour. The vertical grinding first appears in two strength grade rails in the early stare of service.using process and no lateral grinding occurs. With the increase of total weight, the lateral grinding generally appears in both rails. The final rail profiles of 880 MPa-grade and 980 MPa-grade rails are presented in Figures 5 and 6, respectively. It indicates that when the total weight reaches 22.5 million tons, the side grindings of 880 MPa-grade (U71Mn) and 980 MPa-grade(U75V) rails reach 9.85 mm and 3.11mm, respectively. The vertical grindings develops slowly and reach 2.67 mm and 1.43 mm, respectively. Therefore, with the increase of the total weight, the wears of two strength-level rails are dominated by side grinding and the vertical grinding is small. In addition, the vertical grinding and lateral grinding amounts of 880 MPa-grade rail are far greater than those of 980 MPa-grade rail. This is because the wheels and rails mainly contact in the rail side during the train's running on curve section, leading to more side grinding. The wear of 880 MPa-grade rail is more serious because of low strength and low hardness [18].

The injury of rails is divided into slightly damage and serious damage. When the vertical abrasion of rail is >11 mm and the side wear is >19 mm, the rail is classified to serious damage [19]. The seriously damaged rails should be replaced immediately. According to the measurement of rail profiles in Figures 5 and 6, it can be seen that under the same total weight condition, the wear on work side of 880 MPa rail is more serious and crushed deformation happens. The rail profile of 980 MPa rail is mainly characterized by lateral grinding and the crushed deformation doesn't happen. The rails should be replaced when side grinding further increases to 12 mm. Based on the wear trend, 880 MPa rail will be replaced ahead 980 MPa rail. It means that service life of 980 MPA rails is longer under same total weight.

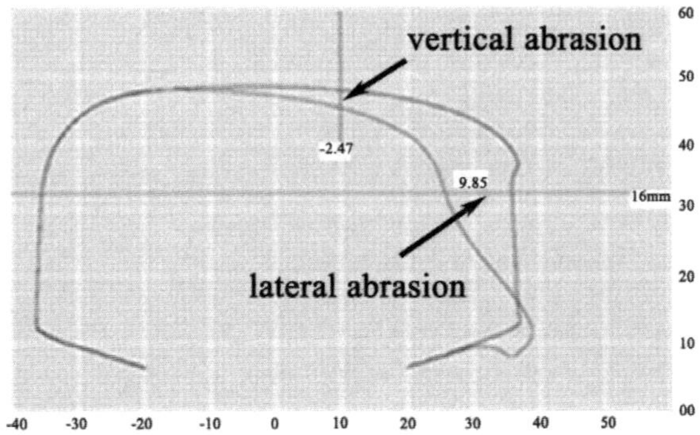

Figure 5. The profile of 880 MPa rail after total weight of 22.5 million tons.

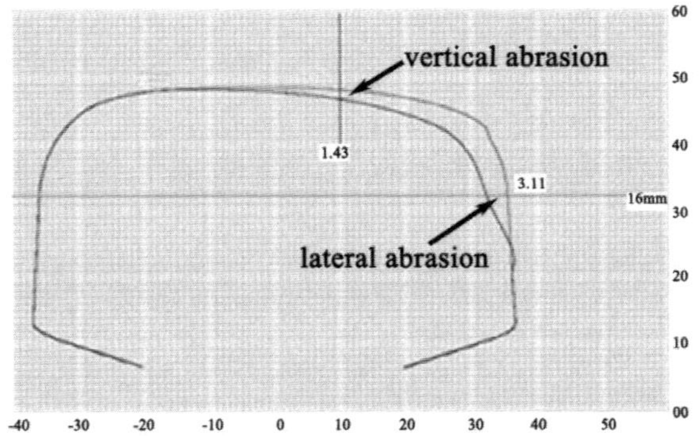

Figure 6. The profile of 980 MPa rail after total weight of 22.5 million tons.

4. Conclusion

The wear tests were conducted by laying the 880 MPa and 980 MPA-grade rails on the cargo line with small curve radius. The influence of total load on wear was analyzed and the following conclusions are obtained:

(1) The service performance data of 880 MPa and 980 MPa-grade rails in the cargo line with small curve radius are analyzed by continuous tracking of service performance. It shows that 880 MPa-grade rail wears faster and side grinding amount is 3 times of that 980 MPa-grade rail. The wear resistance of 980 MPa-grade rails is better, but they are prone to defects of fish scales and striping off.

(2) On the cargo line with small curve radius, 880 MPa-grade rail will be replaced early due to more serious wear. It is recommended to first select 980 MPa-grade rail.

5. Reference

[1] Yang L 2013 *Technol. Forum.* 434-5 (In Chinese)
[2] Yu X M and Feng L C 2013 *Hot working Technol.* **42** 215-7 (In Chinese)
[3] Lin S and Dai H 2016 *Hot working Technol.***45** 252-4 (In Chinese)
[4] Chou J S, Kim C, Tsai P Y, Yeh C P and Son H 2017 *Ksce J. Civil Eng.* 1-16.

[5] Pessel S and Mensinger M 2016 *Transp. Res. Procedia* **14** 2006-2014
[6] Srivastava J P, Kiran M V R, Sarkar P K and Ranjan V 2017 *Procedia Eng.* **173** 1130-1137
[7] Liu L, Pan J Z, Zhou M Y, Chen C H and Zhao X J 2016 *Hot working Technol.* **45** 143-5 (In Chinese)
[8] Zhu W and Ma H 2016 *International Conference on Computer Science and Network Technology* 1498-1501
[9] Ashimov A A, Ashimov A A, Borovski Y V and Volobueva O P 2014 *J. Rail & Rapid Transit* **228** 581-589
[10] Kanematsu Y, Matsui M and Tsuijie M 2017 *QR of RTRI* **58** 50-56
[11] Pan R, Zhao X J, Pan J Z, Liu P T and Ren R M 2014 *Ordnance Mater. Sci. Eng.* **37** 85-6 (In Chinese)
[12] Nippon K H K 2001 *China. Railw. Sci.* **22** 6-15 (In Chinese)
[13] Krylov V V 1996 J. *J. Acoust. Soc. Am.* **100** 3121-34
[14] Yi B 2007 Railw. *Eng.* 91-3 (In Chinese)
[15] Liu Q Y, Wang W J and Zhou Z R 2007 *Lubr. Eng.* **32** 11-14 (In Chinese)
[16] JIANG Z Q, Si D L, Li W and Du X G 2014 *China. Railw. Sci.* **35** 9-10 (In Chinese)
[17] Chang W H 2011 *J. Railw. Eng. Soc.* **2** 71-3 (In Chinese)
[18] Zhou Q Y, Zhang J F, Guo Z W, Xi N S and Gao X P 2010 *China. Railw. Sci.* **3** 27-31 (In Chinese)
[19] Guo Z W 2011 *Railw. Eng.* 106-9 (In Chinese)

IOP Conf. Series: Materials Science and Engineering **244** (2017) 012026 doi:10.1088/1757-899X/244/1/012026

The morphologies of oxide scale in a Si-containing steel

M X Zhou[1], G Xu[1] and Q Yuan[1]

1 The State Key Laboratory of Refractories and Metallurgy, Key Laboratory for Ferrous Metallurgy and Resources Utilization of Ministry of Education, Wuhan University of Science and Technology, Wuhan 430081, China
Guang Xu (Corresponding auther) xuguang@wust.edu.cn

Abstract. Red scale is a kind of strip surface defects which usually appear in Si-containing steels, so that it is important to study the oxidation process of Si-containting steels in order to control this defect. In the prsent study, oxidation epxeriments at different temperatures were conducted for a Si-containing steel, and the morphologies of the oxide scale in the Si-containting steel were investigated. The results show that the oxide scales of the Si-containing steel mainly consist of three layers. The outer layer is Fe_2O_3, the middle layer contains FeO and Fe_3O_4, and the inner layer is composed of of Fe_2SiO_4 and FeO. When the heating tmepraure is above the melting temperature of Fe_2SiO_4 (1173 °C), FeO distributes on Fe_2SiO_4 in the form of dot or lamella and Fe_2SiO_4 shows net-like morphology. However, when the heating temperate is below 1173 °C, no FeO is observed within Fe_2SiO_4 and Fe_2SiO_4 is blocky. In addition, the distribution of Si mainly concentrates in the inner layer and the concentration of oxigen decreases from the outer layer to the inner layer.

1. Introduction

Advanced high strength steel usually contains silicon (Si) to improve its strength [1-6]. Silicon has many beneficial effects on the microstructure and the property of steels. For example, it is well known that Si acts as a role of solid solution strengthening [1]. Besides, Si can hinder the precipitation of carbides, and therefore, a novel carbide-free bainite steel was designed [1, 2]. However, the Si addition may bring some surface defects in steels such as red scale defect. Red scale is mainly composed of Fe_2O_3 which appears on the surface of the hot-rolled strips and deteriorates the surface quality of the strips.

It has been reported that the appearance of red scale is closely related to the addition of Si, so that researchers have paid attention to the oxidation process of Si-containing steels [7-13]. Fukagawa et al. [7] and Okada et al. [8] studied the formation mechanism of red scale. They reported that a phase termed as fayalite (Fe_2SiO_4) forms during the oxidation process in Si-containing steels and it strongly fixes the FeO layer on the steel substrate. The FeO layer, therefore, is difficult to be wiped off and is easily to be oxidized to Fe_2O_3, i.e., red scale defect, during cooling processes after hot rolling. Yang et al. [9] focused on the oxidation kinetics of Si-containing steels and found that the oxidation rate of Si-containing steel decreases with Si addition when the oxidation temperature is below 1000 °C. Some other studies reported the effect of oxygen content on the oxidation process of Si-containing steels [10, 11]. So far, some studies have been performed on the oxidation of Si-containing steels. However, some basic knowledge such as the distribution of Fe_2SiO_4 has not been fully clarified. Therefore, more investigations on the morphology of oxide scale in Si-containing steel should be conducted.

The present study aims to study the morphologies of the oxide scale in a Si containing steel. The effect of heating temperature on the morphology of Fe_2SiO_4 was investigated.

2. Experimental procedure

The compositions of the tested steel are presented in Table 1. Oxidation experiments were conducted in an industrial heating furnace in which the oxidation atmosphere is a combination of 2 vol% oxygen, 13 vol% carbon dioxide, 11 vol % water vapor, and 74 vol % nitrogen. The specimens for oxidation experiments are cuboids with a size of 100×30×10 mm. As shown in Figure 1, the specimens were heated to a certain temperature (1050 °C or 1230 °C) by sectional heating route and then held for 30 min, followed by water cooling to room temperature. Then, the morphology of the oxide scale was observed using a Nova 400 Nano scanning electron microscope (SEM). The composition of the oxide scale was measured by energy-dispersive spectroscopy (EDS). The area scanning and line scanning were also conducted by EDS to analyse the element distribution in the scale.

Table 1. The chemical compositions of the tested steel (wt%)

C	Si	Mn	P	S
0.12	1.52	1.05	0.01	0.01

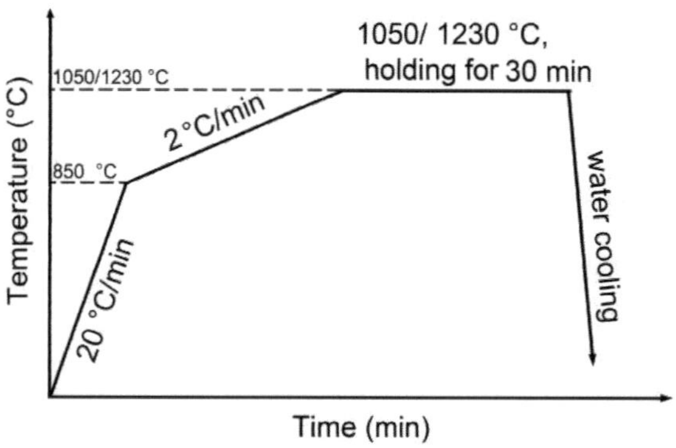

Figure 1. Procedures of oxidation experiments.

3. Results and discussions

Figure 2 shows the morphologies of oxide scale in the tested steel, in which the final heating temperature is 1050 °C. It obviously shows in Figure 2a that the oxide scale mainly consists of three layers. The outer layer and the inner layer account for relative smaller fraction in the whole scale, whereas the middle layer accounts for the largest fraction. Magnifying morphologies of different layers are shown in Figures. 2b~2d, respectively. According to the EDS results (Table 2), the outer layer (Figure 2b) mainly contains Fe_2O_3. The middle layer (Figure 2c) consists of FeO and Fe_3O_4. Some Fe_3O_4 are large irregular blocks and others are small square blocks. The inner layer is composed of FeO and Fe_2SiO_4, where the darker areas are Fe_2SiO_4 which dispersively distributes in the inner layer. Fe_2SiO_4 shows blocky morphology when the final heating temperature is 1050 °C.

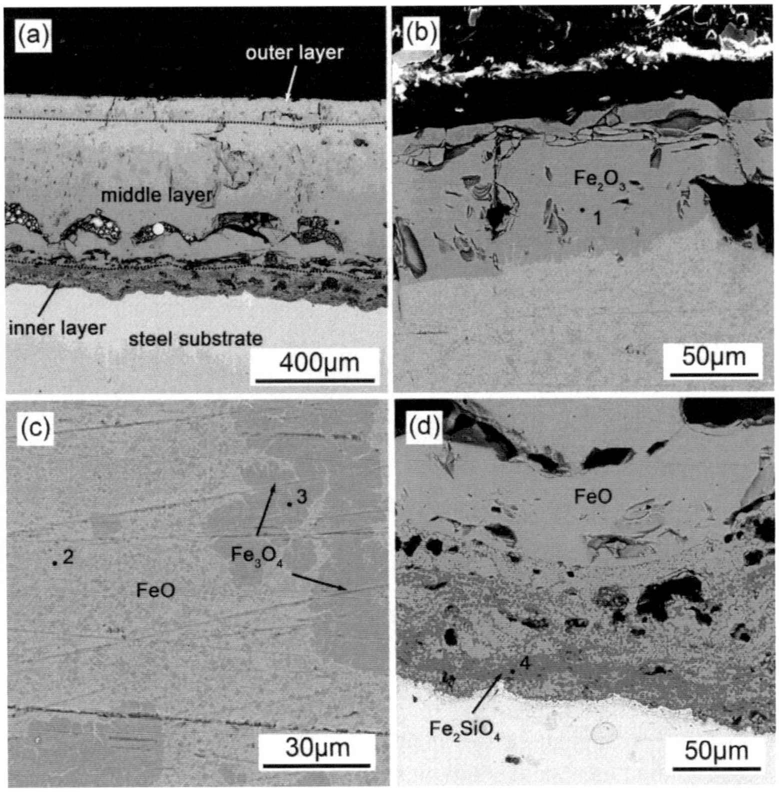

Figure 2. Morphologies of the oxide scale in the tested steel with a final heating temperature of 1050 °C: (a) the whole scale; (b) outer layer; (c) middle layer; and (d) inner layer.

Table 2. The atomic percentages (%) of chemical elements in different layers in the tested steel with a heating temperature of 1050 °C.

Location	O	Fe	Si	Mn
Point 1	60.12	39.67	-	0.21
Point 2	52.33	47.06	-	0.61
Point 3	58.13	41.32	-	0.55
Point 4	57.26	27.28	14.61	0.85

The element distribution in the oxide scale is presented in Figure 3. The area scanning results (Figure 3b and 3c, light areas) show that Si mainly distribute in the inner layer because Fe_2SiO_4 mainly exists in this layer. Line scanning from point A to B (Figure 3d) also confirms that the distribution of Si in the scale mainly concentrates in the inner layer. In addition, area scanning (Figure 3c) shows that oxigen distributes everywhere in the scale (Figure 3c), wheareas line scaning (Figure 3e) presents more detailed result and demonstrates that the concentration of oxigen in the scale decreases from the outer layer to the inner layer. This is consistent with the composition of the oxide scale.

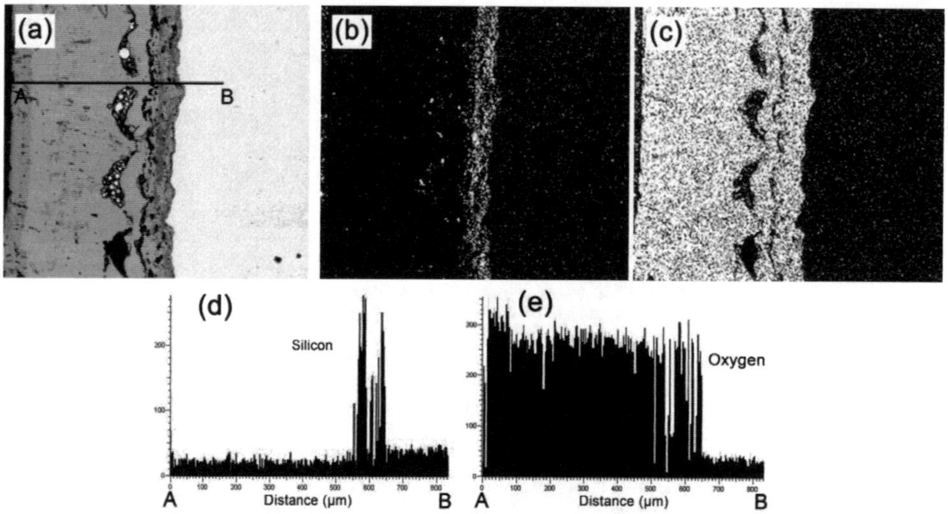

Figure 3. Element distribution in the oxide scale in the tested steel which is heated to 1050 °C: (a) The whole scale; (b) area scanning of silicon; (c) area scanning of oxygen; (d) line scanning of silicon; and (e) line scanning of oxygen.

Figure 4 shows the morphologies of the oxide scale in the tested steel in which the final heating temperature is 1230 °C. It is shown in Figure 3a that the oxide scales also contain three layers when the final heating temperature is higher. However, magnifying figure (Figure 4b) demonstrates that the morphology of Fe_2SiO_4 in the inner scale is obviously different from that in the specimen with a lower heating temperature (1050 °C, Figure 2d). The Fe_2SiO_4 is net-like when the heating temperature is 1230 °C, whereas it shows blocky morphology when the heating temperature is 1050 °C. This difference in morphology is mainly caused by the fact that the melting temperature of Fe_2SiO_4 is about 1173 °C [14]. Fe_2SiO_4 is solid below 1173 °C, and when the temperature exceeds 1173 °C, solid Fe_2SiO_4 melts and it penetrates into outer place along the narrow gaps in the scale. As a result, net-like morphology forms. Compared to the blocky morphology of Fe_2SiO_4, net-like morphology is more harmful because net-like Fe_2SiO_4 can firmly bonds the steel substrate and the FeO. Thus, FeO is difficult to be wiped off during the descaling and rolling processes.

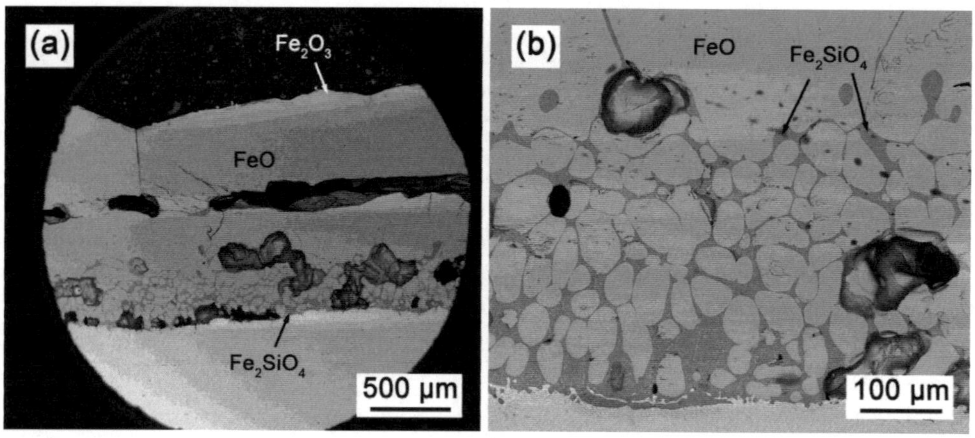

Figure 4. Morphologies of the oxide scale in the 1.52% wt% Si steels with a heating temperature of 1230 °C: (a) the whole scale; (b) inner layer.

The area percentages of each layer in different specimens are calculated and ginven in Table 3. The middle layer, which mainly contains FeO, accounts for the largest fraction. In addition, for the same steel, when the heating temerpature increases from 1050 °C to 1230 °C, the fraction of inner layer increases because liquid Fe_2SiO_4 can penetrate into outer place and account for larger fraction.

Table 3. Area percentages (%) of each layer in different specimens.

Specimen	Outer layer	Middle layer	Inner layer
1230 °C	3.1	72.6	24.3
1050 °C	11.5	75.0	13.5

The magnifying morphologies of the Fe_2SiO_4 is shown in Figure 5. It is interesting to note that when the heating temperature is higher than 1173 °C (Figure 5a), FeO distributes on Fe_2SiO_4 in the form of dot or lamella. However, when the temperature is below 1173 °C, the surface of Fe_2SiO_4 is fresh and there is no FeO observed within Fe_2SiO_4. Therefore, whether there is FeO within Fe_2SiO_4 depends on the heating temperature.

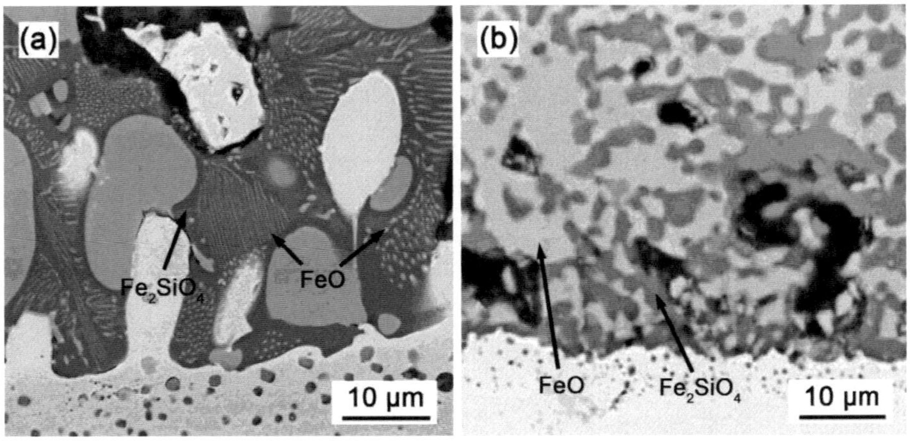

Figure 5. Morphologies of Fe2SiO4: (a) 1230 °C; (b) 1050 °C.

4. Conclusions

Industrial oxidation epxeriments were conducted to investigate the morphologies of oxide scale in Si-containting steels. The effect of heating temperature on the morphology of oxide scale, especcially Fe_2SiO_4, was investigated. The results show that the oxide scales of Si-containing steels mainly consist of three layers. The outer layer is Fe_2O_3, the middle layer contains FeO and Fe_3O_4, and the inner layer is composed of Fe_2SiO_4 and FeO. In addition, when the heating tmepraure is above 1173 °C (the melting temperature of Fe_2SiO_4), FeO distributes on Fe_2SiO_4 in the form of dot or lamella and Fe_2SiO_4 shows net-like morphology. However, when the heating temperate is below 1173 °C, no FeO is observed within Fe_2SiO_4 and Fe_2SiO_4 is blocky. Moreover, the distribution of Si mainly concentrates in the inner layer and the the concentration of oxigen decreases from the outer layer to the inner layer.

5. References

[1] Caballero F G and Bhadeshia H K D H 2004 *Curr. Opin. Solid State Mater* **8** 251
[2] Caballero F G, Bhadeshia H K D H, Mawella K J A, Jones D G and Brown P 2001 *Mater. Sci. Technol* **17** 512
[3] Zhou M X, Xu G, Wang L and Hu H J 2016 *Met. Mater. Int.* **22** 956
[4] Zhou M X, Xu G, Tian J Y, Hu H J and Yuan Q 2017 *Metals* **7** 263
[5] Hu H J, Xu G, Zhou M X and Yuan Q 2016 *Metals* **6** 173
[6] Tian J Y, Xu G, Zhou M X, Hu H J and Wan X L 2017 *Metals* **7** 40
[7] Fukagawa T, Okada H and Maehara Y 1994 *ISIJ Int.* **34** 904
[8] Okada H, Fukagawa T and Ishihara H 1995 *ISIJ Int.* **35** 886
[9] Yang Y Y, Yang C H, Lin S N, Chen C H and Tsai W T 2008 *Mater. Chem. Phys.* **112** 566

[10] Abuluwefa H, Guthrie R I L and Ajersch F 1996 *Oxid. Met.* **46** 423
[11] Chen R Y and Yuen W Y D 2003 *Oxid. Met.* **59** 433
[12] Yuan Q, Xu G, Zhou M X, He B and Hu H J 2017 *Metals* **7** 37
[13] Fukagawa T, Okada H and Maehara Y 1995 *ISIJ Int.* **81** 559
[14] Liu X J, Cao G M, He Y Q, Jia T and Liu Z Y 2013 *J. Iron Steel Res. Int.* **20** 73

Acknowledgments
The authors gratefully acknowledge the financial supports from National Natural Science Foundation of China (NSFC) (No. 51274154) and the Special Fund of Wuhan University of Science and Technology for Master Students' Short-Term Studying Abroad.

IOP Conf. Series: Materials Science and Engineering **244** (2017) 012027 doi:10.1088/1757-899X/244/1/012027

Industrial wastewater treatment using higher aquatic vegetation in the former mining company of the Far Eastern Federal district

L. T. Krupskaya[1,2,5], V. P. Zvereva[1,3,4], K. E. Gula[1], L. P. Gul'[1,2], D. A. Golubev[1,2] and M. Yu. Filatova[1,2]

1 Pacific National University, 136, Tikhookeanskaya str, Khabarovsk, 680035, Russian Federation
2 Far East Scientific-Research Institute of Forestry, 71, Volochaevskaya str, Khabarovsk, 680020, Russian Federation
3 Far East Federal University, 8, Sukhanova str, Vladivostok, 690091, Russian Federation
4 Far East Geological Institute FEB RAS, 159, Prospect 100 let Vladivostoka, Vladivostok, 690022, Russian Federation
5 ecologiya2010@yandex.ru

Abstract. The article describes the results of studying the problems of industrial wastewater treatment using higher aquatic vegetation (hydrophytes) in the former mining enterprise of the Far Eastern Federal District (FEFD). They are aimed at reducing the negative environment impact of toxic tin ore wastes. The material of research were drainage, mine and slime waters as well as Lemna minor and Common reed grass (Phragmites communis). In the work conventional modern physico-chemical, chemical, biological and mathematical-statistical methods were used, as well as in the process of research the methods of atomic absorption spectrophotometry for AAS and mass spectrometry with inductively coupled plasma on ISP-MS ELASN DRS II PerkinElmer was applied.
The data obtained in the course of the experiment (2015-2016), indicate that a degree of wastewater treatment, using Lemna minor, is high. Virtually, all compounds of toxic chemical elements contained in industrial wastewater (zinc, cobalt, nickel, cadmium, iron, manganese, lead, etc.) were fully absorbed by a hydrophyte. Pollutant extraction was almost 95%. The obtained results of the study in laboratory conditions proved the possibility of effective use of the Lemna minor for the purification of drainage and mine waters.
A key contribution of this paper is the relationship between possible toxic metals contained in industrial wastewater and a higher degree of absorption by their higher aquatic vegetation. These hydrophytes absorb these possible toxic metals in an aqueous medium and are contaminated with these heavy metals.

1. Introduction

Intensive development of mineral resources in the last century in the Far Eastern Federal District (FEFD) led to the formation of environment- mining systems. They comprise tin ore enrichment waste stored in the tailings, as well as in sedimentation tanks, dumps, drainage, mining and slurry waters. It is known that industrial waste discharge into water contains large amounts of pollutants with carcinogenic, general toxic, cumulative and mutagenic activity and pose a threat to the ecological balance of natural systems. Therefore, the protection of air, soil, biota, surface and underground water from pollution by toxic heavy metals is an important environmental challenge [1, 2, 3, 7].

IOP Conf. Series: Materials Science and Engineering **244** (2017) 012027 doi:10.1088/1757-899X/244/1/012027

Environmental and social security for all living in the study area should be ensured in the near future by improvements in the technology of industrial wastewater treatment. However, in the former mining companies of FEFD this problem was not studied virtually. In this regard, the following goal has been defined: to carry out pilot studies at industrial water treatment plant using biological systems (higher aquatic vegetation, hydrophytes) to reduce their negative impact on environment. Accordingly, the following goals were formulated: 1. The analysis, generalization and systematization of literary sources on the above problem; 2.The experiment on treatment of drainage, mine and slime water using Lemna minor; 3.The proposals to reduce the negative impact of effluents from the former mining enterprise on the environment.

2. Materials and methods of research

In the work conventional modern physico-chemical, chemical, biological and mathematical-statistical methods were used, as well as in the process of research the methods of atomic absorption spectrophotometry for AAS and mass spectrometry with inductively coupled plasma on ISP-MS ELASN DRS II PerkinElmer was applied.

3. Results and discussion

The analysis, generalization and systematization of the literature data for the industrial water treatment of former mining enterprises in Russia shows that the researchers authentically showed the active role of microorganisms, algae, aquatic vegetation in biological treatment of heavy metals and organic pollutants [5, 6, 3]. However, these problems have not been practically studied for conditions of mining enterprises in FEFD, there is only some experience [8].

As a result of tin ore processing at the tin concentration plant the tin waste (pulp, industrial waste water, sludge) containing polluting substances (ions of heavy non-ferrous metals, suspended solids of mineral origin, etc.) are formed, they constitute an ecological threat to environmental components.

Industrial waste water treatment is a complex of measures for removing impurities contained in industrial waste waters. The treatment at operating tin extraction plants was carried out by mechanical and physico-chemical methods.

However, at the end of the last century, bankruptcy of the former Khrustalnensky GOK (Primorsky Krai) resulted in liquidation of hydro-engineering service. The tailings and ore processing plant were left to fend for themselves. In the Kavalerovsky district the main and only production element was Sn, and sulphides containing Cu, Pb and Zn in industrial quantities were always left in tails and dumped in tailings. Almost all reagents used at plant (cyanide compounds, xanthates, phenols, lime, acid, liquid glass, pine oil and other substances) used in the process together with the pulp got into tailings. During accidents at mining and processing plant, a breakthrough of the dam during snow melting and heavy rain the size of slime lakes grew and tailings discharge and slime water draining in river network occurred. In this region similar situations on tailings ponds were observed quite frequently. So at the end of 1960, in Kavalerovsky district the slimy lake and virtually all the accumulated tailings were washed away in the Zerkalnaya river. When slime and drainage waters got into into surface and groundwater, they led not only to a change in their chemical composition, but caused the death of living organisms.

Hypergene processes in tailing pond of the former "Khrustalnensky GOK" occur both on the surface and in tails mass. From the hypergenesis zone and technogenic system a wide range of chemical elements is carried out by mine, drainage, slime and interstitial water into surface and groundwaters, resulting in change of geochemical background .

The evidence of Khrustalnesky GOK operated in the last century (1988) shows that the liquid effluents of mine waters only in 1988 amounted (kg): Cu - 66.1, Zn -4178.8 and Fe -1104.5. Mine debit of dewatering was unsteady, significantly fluctuated, reaching sometimes 3600 m^3 per day. Pumped out water was usually discharged to dump waste rock and then draining through it, came to surface watercourses, and only during the period of exploitation it partially passed through the tanks. Consequently, during decades mining waters not restrained for 24 hours carried large amounts of pollutants into ecosphere, including water bodies. Our monitoring of hydrochemical samples of waters in slime lakes, drainage, mine, and from the springs below the tailings indicate that hydrochemical samples selected during the operation of the mining and concentration plant (2001 and 2003), and after its shutdown (2006-2016) are intensely enriched with compounds of toxic heavy metals and elements, and they are permanently carried out beyond technogenic systems, contaminating not only surface and groundwater but also river water [7].

The studies showed that in Kavalerovsky district of FEFD the mining industrial complexes with pronounced specificity of negative impact on the environment formed. These areas are characterized by tense and crisis ecological situation. The content of Zn, Cu, Pb, Cd, Fe, Mn, Al, As compounds and other elements in mine, drainage, slime and river waters during 2001-2016 frame in dosens, hundreds and thousands of times are above the background characteristics in "reference" fresh waters. The dilution of these waters should be in hundreds, thousands and even tens of thousands of times, which is practically impossible. Due to the destruction of aquatic biocoenoses self-purification of waters is slower than the flow of contaminated mine drainage. This leads to the destruction of the aquatic ecosystems, changing the hydrochemical background of the area and the quality of natural waters for the worse.

The tendency to increase their concentration in time is observed for a number of chemicals under consideration. The high activity of hypergene processes in technogenic systems in the past century and at present has resulted in significant contamination of the ecosphere in general. The authors assume that the future status of the ecosphere will change for the worse, mainly as the sulfides are oxidized over the centuries. It should be noted that the indigenous reserves of ores in this area have not been fully developed, and the tailings are actually complex technogenic deposits of a wide range of Sn, Cu, Pb, Zn, Cd, In and other elements.

The accumulated in the last century large amount of pollutants as a result of the activities of the former Khustalnensky GOK, including industrial waste waters and their export into ecosphere, establishes the need for their treatment, because the technological waste water is among the most environmentally hazardous wastes. Technogenic pollution brings also a great economic damage to the environment objects. All this testifies to the need to undertake a number of environmental activities as soon as possible. In this regard, the authors continued experimental studies of the treatment of waste water of this former mining enterprise using Lemna minor. The experiments have been started in 2015 (Figures. 1 a, b, c, 2 a, b).

In the laboratory, during 2015-2016. (July 7– September 6), the following experiment was conducted: vegetation vessels of 4 litres volume were filled with the samples in three repeatabilities: 1) water from the Fabrichny sediment pond tailings of the former Khrustalnensky GOK; 2) water from the Vysokogorsky sediment pond tailings of the former Khrustalnensky GOK. At the beginning of July Lemna minor was planted in them.

It is known that higher aquatic plants (hydrophytes) by the biomass of the living matter in water bodies occupy a leading position as absorbents of pollutants.

For this purpose the plant species with the highest enzyme activity and rate of biodegradation, with good growth and adaptation in these climatic conditions, as well as vegetative seasonally (Lemna minor and Phragmites communis) were selected. In addition, these plants possess useful properties (rhizofiltration, phytoextraction, phytodegradation, phytostimulation) (Figures. 1 a, b, 2).

The basis of treatment technology is the ability of higher aquatic vegetation (HAV) to extract contaminants from industrial wastewater: arsenic compounds, zinc, copper, lead, nickel, cobalt, manganese and cyanide (detoxification) by biodegradation, bio-oxidation and sorption from solutions by hydrophyte plants Lemna minor and Phragmites communis in severe climatic conditions. Furthermore, the aquatic plants have and other basic functions (filtration, absorption, accumulation, sanitary, oxidative) in water basins, that significantly improves the quality indicators of industrial waste water.

In the process of studying it was found that Lemna minor and Phragmites communis have highly absorptive capacity, are characterized by stability and ruggedness to changes in water hydrology and hydrochemistry, have a wide amplitude of distribution in the surrounding landscape, have sustainability and ability to grow in the industrial environment in different types of waste water, and extracting pollutants are easily removed from the basins not accumulating silt and providing sufficient

IOP Conf. Series: Materials Science and Engineering **244** (2017) 012027 doi:10.1088/1757-899X/244/1/012027

aeration of the water, destroying, pathogenic bacteria. The vegetation (hydrophytes) collected after the high biotreatment contains useful product (tin and other chemical elements).

Selection of the studied samples for chemical analysis (Figure 1, 2) was made from each container for the identification of arsenic, zinc, copper, lead, iron, nickel, cobalt, manganese and cyanide. The ambient humidity, the environment temperature and pH of the analyzed samples were determined.

Bellow, the results of the analyses of the change of chemical compounds concentration in the technological wastewater of Fabrichny tailings in the course of the experiment are shown (Figure 3).

Figure 1. Experimental studies of the industrial wastewater treatment of the former “Khrustalnensky GOK" ("Fabrichny tailings") using Lemna minor (beginning of the experiment: a- side view, b-top view)

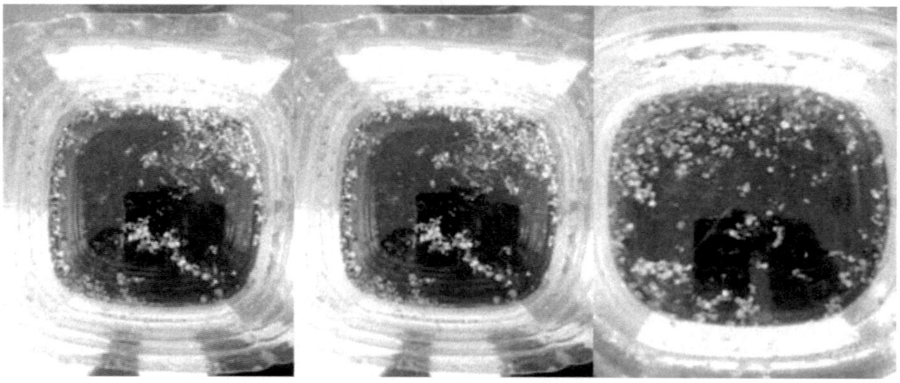

Figure 2. Finishing of the experimental studies of the industrial wastewater treatment of the former "Khrustalnensky GOK" ("Fabrichny tailings") using Lemna minor (top view)

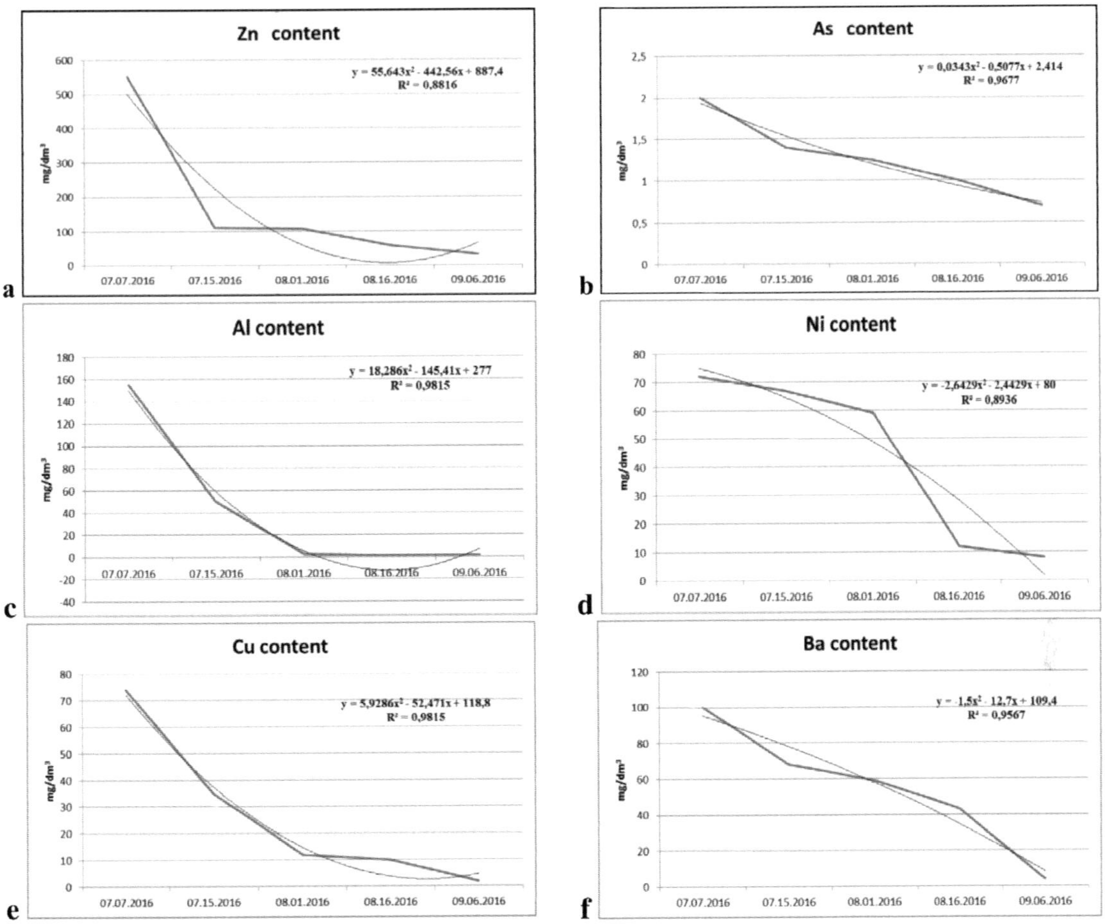

Figure 3. Change of Zn (a), As (b), Al (c), Ni (d), Cu (e) and Ba (f) compounds concentrations in the industrial wastewater from the former "Khrustalnensky GOK" ("Fabrichny"tailings) in the course of the experiment

Table 1 presents the results of polluting substances content in Lemna minor at the end of the experiment

Table 1. Content of polluting substances in Lemna minor the end of the experiment with technological effluents from the tailings of the former mining enterprise "Khustalnensky GOK"

Name	Mo	Co	Ni	Cu	Zn	Cd	Ba	W	Hg
Lemna minor, Control	1.32	8.37	8.50	25.65	185.47	0.54	238.30	2.43	0.19
Lemna minor, water from tail. , variant 1	1.96	20.16	158.44	148.01	1264.84	3.40	333.26	5.73	0.27
Lemna minor, water from tail., variant 2	2.41	21.09	181.30	149.01	1308.14	2.48	346.93	6.53	0.49
Lemna minor, water from tail., variant 3	2.11	17.32	117.44	107.80	835.66	2.15	309.98	3.88	0.26

The data obtained in the course of the experiment (2015-2016) suggest that the degree of industrial waste water treatment using Lemna minor is high. Virtually, all toxic chemicals contained in industrial wastewater (zinc, cobalt, nickel, cadmium, iron, manganese, lead, etc.) were fully absorbed by a hydrophyte. Almost 95% of contaminants were extracted.

4. Conclusion

The results of the research in the laboratory have proven an effective use of Lemna minor for the purification of drainage and mine waters.

5. References

[1] L.M. Kondratyeva, V.V. Bardyuk and A. G. Zhukov: Accumulation and transformation of toxic substancesin ice of Amur and Songhua rivers after technogenic accident in China in 2005. in Led i Sneg (Ice and Snow) No.4 (2011), p. 118-124

[2] O.L. Gas'kova: Aspects of soil contamination in the area of waste storage of pyrometallurgical zinc extraction: in O.L. Gaskova, S.B. Bortnikova, V.G. Kabanik and S.P. Novikova. Chemistry for sustainable development, vol. 20, No 4 (2012) p. 419-429

[3] L.T. Krupskaya, A.V. Leonenko, V.P. Zvereva *Impact of technogenic systems on the environment and human health in the Priamurye and Primorye territories*: in Contemporary Problems of Ecology. Vol. 6, No 2 (2013) p. 223-227

[4] L.T. Krupskaya, M.B. Bubnova, V.P. Zvereva and A.V. Krupskiy: Characteristics of mining-ecological monitoring of environmental objects changing under the influence of toxic waste tailing dump ("Solnechny GOK" Company) in Environmental Monitoring and Assessment. Vol. 184, No 7 (2012) . p. 2775-2781

[5] R.I. Rozvaga: On the mechanism of biological wastewater and cyanic pulp treatment from metal cyanides and rhodonides: in R.I. Rozvaga, A. N. Klets and S.S. Nurkeev. Ores and metals No 1(1997), p. 14-15

[6] S.S. Timofeeva. Ecological biotechnology in: S.S. Timofeeva. Ircuysk: IGTU. 1999. 210 p

[7] V.P. Zvereva and L.T. Krupskaya. Rare earth elements in mine, slime, and river waters in the Kavalerovsky and Dalnegorsky districts of the Russian Far East: Russian Journal of General Chemistry. Vol. 85. No 13 (2015) p. 2867-2873

[8] K.E. Gula, O.O. Panfilov, L. Y. Krupskaya. Treatment of industrial tin ore wastewater using hydrophytes in the Far Eastern Federal District: in Ecological present problems, XV International Research Conference (2016) p.67-69

Acknowledgments

The investigation is fulfiled at the expense of the grant of the Russian scientific fund (Project № 15-17-10016), Pacific National University.